플라스틱 사출금형 설계의 기본과 노하우

한성렬, 박태원

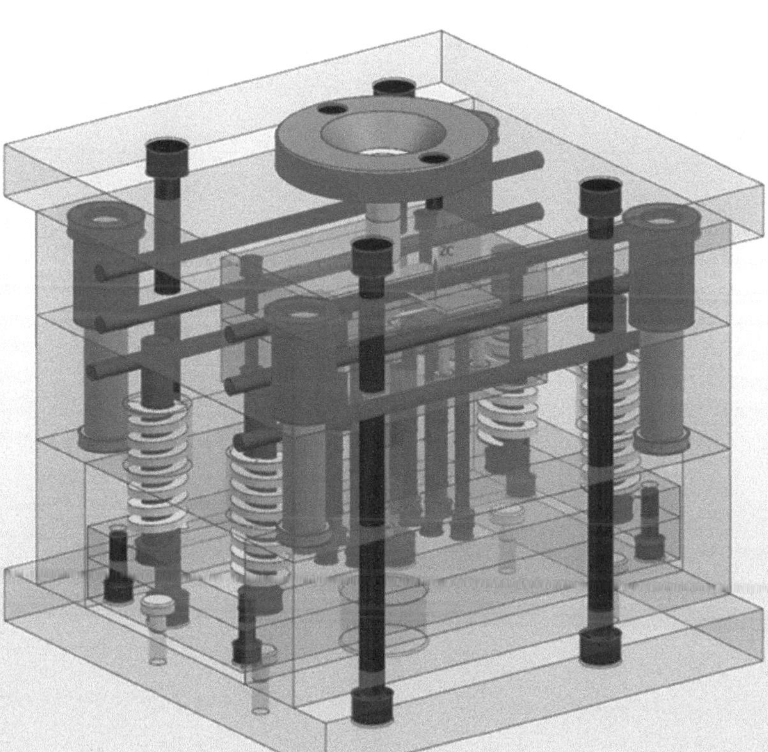

도서출판 WisdomPL

머 리 말

플라스틱 금형 산업은 세계 경제 침체의 영향을 받아 전체적으로 정체 상태에 빠져 있다. 그러나 활동 분야에 따라 매우 견고한 산업 유형도 있다. 더구나, 이 세계에 약 70억 명의 인구가 살고 있다. 우리의 눈이 전 세계로 향하게 되면, 우리는 플라스틱 제품에 대한 수요가 엄청나게 크다는 것을 알게 된다.

현재와 같은 어려운 경제 변동 속에서도 매우 바쁜 금형 제조업체가 있다. 이 회사들은 모두 공통점을 가지고 있다. 즉, 이 회사는 이전과는 달리 연구 및 기술 개발을 꾸준히, 그리고 확실히 수행해 왔다. 기술 개발과 연구에는 많은 돈이 필요하다고 생각되지만, 소액 또는 대규모의 돈이 지출되었는지 여부에 관계없이 중요한 점은 주의를 기울인 주제가 옳은 것인지 또는 틀린 것인지에 관한 것이다.

요즈음 대중적으로 이용 가능한 정보 및 기술을 사용하여 금형의 설계 및 제조를 90%까지 쉽게 달성 할 수 있지만 100%에 가까워지는 접근법은 기술의 차이다. 나머지 10%를 충족시킬 수 있는 힘은 회사의 경쟁력이다. 최대 90%까지 도달하는 속도에 대해 경쟁하는 것만으로는 충분하지 않다. 우리는 당신이 그 최종 10%의 완성을 빠르고 정확하게 할 수 있는 전문가가 되길 원한다.

기술을 향상시키는 것은 신뢰성이 높은 기술 정보를 수집하는 방법에 달려 있다. 프로토 타입 준비 및 실험이 전형적인 예이다. 또한 많은 기술 정보 중에서 실제 정보를 찾아 낼 수 있는 능력을 키우는 것이 매우 중요하다.

(1) 성형품의 품질을 보장 할 수 있는 금형

(2) 높은 생산 능력을 제공하는 금형

(3) 유지보수하기 쉬운 금형

(4) 특수 성형품을 생산 할 수 있는 금형

위와 같은 요구 사항을 충족시키는 금형은 세계 시장에서 매우 높게 평가되며 가격도 적절하게 높다.

좋은 외국 회사의 주문을 받으려면 외국 고객 회사의 금형설계자 또는 관리자와 영어 또는 다른 언어로 대화하고 원활하게 의견을 교환해야한다.

이것이 기회라고 생각하지 않는가? 자유 시간을 활용하면 1 년에 약 200 시간을 보낼 수 있다. 그러한 노력을 하는 방법이 새로운 전망을 열어 줄 수 있다

전문가와 세련된 금형 기술을 사용하여 전 세계 사람들의 행복한 삶을 지원하는 제품을 계속 생산 할 수 있기를 바란다.

It is clear that all the technical information is the material that came from MISUMI Corporation.
모든 기술자료는 ㈜MISUMI로부터 나온 자료임을 밝힌다.

차 례

C. 성형방법의 종류 ··· 61

D. 성형불량과 대책 ··· 81

E. 플라스틱수지 특성 ··· 107

G. 금형부품설계 ························ 153

H. 금형제작 ··· 237

I. 유지보수 ··· 267

J. 기타 ··· 285

A. 사출성형기

A 001 Basic Structure of an Injection Molding Machine

An injection molding machine is a machine for converting the plastic material into a liquid state, injecting into the cavity of the mold, opening the mold, and taking out the solidified molded article. A large number of manufacturers in the major industrial countries of the world today are manufacturing and selling various types of injection molding machines, all of which are not being manufactured to any standardized specifications, but different types are being manufactured according to the purpose of use, size of the machine, accuracy, etc. The common types of injection molding machines that are in widespread use in the market have the following structures.

1) Material used: Thermoplastic resin
2) Plasticizing - injection structure: Screw in-line type
3) Power source: Hydraulic type, electric motor driven type, hybrid type (hydraulic and electric motor drives are used together)

In general, an injection molding machine is constituted from the following functional units.

A-001 사출성형기의 기본 구조

사출성형기는 플라스틱 재료를 액상으로 변환하고, 금형의 캐비티에 주입하며, 금형을 열고, 고화된 성형품을 꺼내기 위한 기계이다. 세계의 주요 제조국의 많은 업체들이 현재 비표준화 된 규격으로 다양한 유형의 사출성형기를 제조 및 판매하고 있지만, 사용 용도, 기계의 크기, 정밀도 등에 따라서 제작되고 있다. 시장에서 널리 사용되는 사출성형기의 일반적인 유형은 다음과 같은 구조를 갖는다.

1) 사용 재료 : 열가소성 수지
2) 가소화 - 사출 구조 : 스크류 인라인 타입
3) 구동력 : 유압 타입, 전동 타입, 하이브리드 타입(유압과 전기 모터를 함께 사용)

일반적으로 사출성형기는 아래와 같은 장치들로 구성된다.

1. Plasticizing Unit

The raw materials of the common thermoplastic resin are sold in the market in the form of pellets, and to carry out injection molding, it is necessary to heat and melt the pellets and to convert them into a fluid state. The unit used for this purpose is the plasticizing unit. Plasticizing is melting the material so that it is soft and can be deformed easily in the plastic state. The following structures are used most commonly for a plasticizing unit.

1) Plunger type
2) Inline screw type
3) Pre-plunger type
4) Screw pre-plunger type

In all methods, pellets are loaded into a cylinder and plasticized by heating using an electric heater surrounding the cylinder.

The method of supplying the pellets to the heated part can be of the plunger type or of the screw type. The plunger type has a structure in which a plunger moves forward and supplies the pellets to the heated part. In contrast with this, the screw type has a structure in which a screw with a helical groove is rotated thereby making the pellets move forward through the helical groove and are supplied to the heated part. These two methods have their own advantages and disadvantages, and the selection is made based on factors such as the material, molded article, cost of investment, etc.

1. 가소화 장치

일반적인 열가소성 수지는 펠릿 형태로 시판되고 있으며, 사출성형을 위해서 펠릿을 가열 및 용융시켜 유체 상태로 전환시킬 필요가 있다. 이를 위하여 사용되는 장치가 가소화 장치이다. 가소화는 재료가 용융되어 부드럽고 쉽게 형상이 변하는 상태이다. 다음은 가장 일반적으로 사용되는 가소화 장치들이다.

1) 플런저 타입
2) 인라인 스크류 타입
3) 프리 - 플런저 타입
4) 스크류 프리 - 플런저 타입

모든 타입에서 펠릿은 실린더에 투입되고, 실린더를 에워싸는 전기 히터를 사용하여 가열하므로 가소화된다.

가열 된 부분에 펠릿을 공급하는 방법으로는 플런저 타입과 스크류 타입이 있다. 플런저 타입은 플런저가 전진하여 가열 된 부분에 펠릿을 공급하는 구조를 갖는다. 이와는 다르게 스크류 타입은 나선형 홈이 형성된 스크류가 회전되어 펠릿이 나선형 홈을 통해 전방으로 이동하여 가열 된 부분에 재료가 공급되는 구조를 갖는다. 이 두 가지 방법은 장단점이 있으며 재료, 성형품, 투자비용 등의 요인에 따라 선택된다.

The parameter that needs to be studied in a plasticizing unit is the plasticizing capacity. The plasticizing capacity is the volume of injectable molten plastic resin that can be prepared in unit time, and this affects the volume of the molded article that can be prepared. Therefore, in the case of a molded item that requires a large volume of plastic, it is necessary to prepare a plasticizing unit having a large plasticizing capacity.

가소화 장치에서 고려해야 할 항목은 가소화 용량이다. 가소화 용량은 단위 시간 내에 용융 할 수 있는 수지의 부피이며, 이것은 성형될 수 있는 성형품의 체적에 영향을 미친다. 따라서 다량의 플라스틱을 필요로 하는 성형품의 경우, 큰 가소화 용량을 갖는 가소화 장치를 준비해야 한다.

A 002 Specifications of Plastic Injection Molding Machines

A-002 사출성형기의 사양

Although it is common that the specifications of injection molding machines differ depending on the manufacturer, it is important for practical application to compare the size or dimensions and capacity of machines of different manufacturers.

It is good to check the following specifications while comparing the specifications of different injection molding machines.

사출성형기의 사양은 제조사마다 다르지만 실제로는 제조사마다 사출성형기의 크기나 치수 및 용량을 비교하는 것이 중요한다.

다른 회사의 사출성형기 사양을 비교하면서 다음 사양을 확인하는 것이 좋다.

- Machine name
- Screw diameter
- Injection volume
- Plasticizing capacity
- Injection pressure
- Injection ratio
- Mold clamping force

- 성형기 명칭
- 스크류 직경
- 사출량
- 가소화 용량
- 사출압력
- 사출율
- 형체력

- Mold clamping method
- Mold clamping stroke
- Tie bar distance
- Daylight
- Die plate dimensions
- Machine dimensions
- Machine weight
- Drive method
- Nozzle diameter
- Nozzle tip SR
- Ejector rod diameter
- Type of associated take out machine
- Year, month, and date of purchase
- Cumulative number of shots
- Special specifications

If these specifications are entered in a table using a spreadsheet software such as Excel, at the time of designing a mold, it is becomes easy to organize the information in one's mind and to make appropriate decisions.

Since there are many items that need to be judged during the design of a mold, saving the time and effort required for "searching for information" and using that saved time for making judgments is a skilful method of using time.

- 형체 방법
- 형체 스트로크
- 타이바 간격
- 데이라이트
- 다이 플레이트 치수
- 성형기 치수
- 성형기 중량
- 운전 방식
- 노즐 직경
- 노즐 팁 SR
- 이젝터 로드 직경
- 취출 기기의 유형
- 구입 년, 월, 일
- 누적 성형 횟수
- 특수 사양

Excel과 같은 스프레드 시트 소프트웨어를 사용하여 이러한 사양을 테이블에 입력하면 금형설계 시 정보를 쉽게 구성하고 적절한 결정을 내릴 수 있다.

금형의 설계 과정에서 판단해야 할 항목이 많으므로 "정보 검색"에 필요한 시간과 노력을 절약하고, 그 절약된 시간을 판단을 위해서 사용하는 것은 시간 사용의 기술적 방법이다.

A 003 Basic Knowledge on the Mold Clamping Force

When a injection mold is fixed in a molding machine and molten plastic is injected into the interior of the cavity from the injection nozzle, a high filling pressure acts on the inside of the cavity. Since the parting surfaces of the mold try to expand outward due to this pressure, it is necessary to clamp the mold so that it does not open instantaneously.

It is easy to imagine that flash will be generated if the parting surfaces open even very slightly. The force of keeping the mold closed tightly is called the "required mold clamping force". The unit for the required mold clamping force is N (Newtons), or kgf. tf

At the time of designing a new mold, it is necessary to obtain by theoretical calculations what is the optimum required mold clamping force that the injection molding machine has to have for the mold to be installed in it.

For example, if a required mold clamping force of 100 tf was obtained by calculations, if this mold is installed in an injection molding machine with a 75 tf capacity, the molded product will be full of flash there by making it impossible to carry out the molding operation.

Further, if the mold is installed in a molding machine with a 300 tf capacity, even if the molding operation is possible, since usually the hourly cost of a 300 tf machine is higher than that of a 100 tf machine, the molding operation becomes high in cost.

A-003 형체력에 대한 기본 지식

사출성형기에 금형을 고정하고, 성형기 노즐로부터 용융수지를 캐비티 내부에 주입하면 캐비티 내부에 높은 충전압력이 작용한다. 금형의 파팅면이 그 압력으로 인해 외부로 팽창하려고 하기 때문에 순간적으로 금형이 열리지 않도록 금형을 죄어야 한다.

파팅면이 아주 조금이라도 열리면 플래시가 발생한다고 생각하기 쉽다. 금형을 견고히 닫은 채로 유지하는 힘을 "요구 형체력"이라고 한다. 요구되는 형체력의 단위는 N (뉴톤) 또는 kgf, tf 이다.

새로운 금형을 설계 할 때, 금형을 설치하기 위하여 사출성형기가 가져야하는 최적의 형체력을 이론적으로 계산해야 한다.

예를 들어 계산에 의해 100 톤의 형체력이 필요한 경우, 이 금형을 75톤 용량의 사출성형기에 설치하면 성형품이 플래시로 가득 차서 성형 작업의 수행이 불가능 하게 될 것이다.

또한, 300톤 용량의 성형기에 금형을 설치하면, 성형 작업이 가능하더라도, 보통 300톤 성형기의 시간당 비용이 100톤 성형기의 시간당 비용보다 높기 때문에 성형 비용이 상승한다.

The required mold clamping force of a mold can be calculated using the following equation.

금형의 필요한 형체력은 다음 식을 사용하여 계산할 수 있다.

$F = p \times A/1000$, where,

F: Required mold clamping force (tf),

p: pressure inside the cavity (kgf/cm^2),

A: total projection area (cm^2)

$F = p \times A / 1000$ 여기서,

F : 형체력(ton),

p : 캐비티 내부의 압력 (kgf/cm^2),

A : 총 투영 면적 (cm^2)

Here, p will have a value in the range of 300 to 500 (kgf/cm^2). The value of p varies depending on the type of plastic, molded item wall thickness, cavity surface temperature, molding conditions, etc. To be more accurate, it is recommended to incorporate a pressure sensor inside the cavity, and to collect guideline data from actual measured values. Also, A is the total projection area of the cavity and the runner with respect to the parting surface. Therefore, the value of A varies depending on the number of items molded and on the placement of the runner.

여기서, p는 300 내지 500 (kgf/cm^2) 범위의 값을 가지며, p 값은 플라스틱 종류, 성형품 살 두께, 캐비티 표면 온도, 성형 조건 등에 따라 다르다. 더 정확히 계산하려면 캐비티 내부에 압력 센서를 설치하고 실제 측정 값에서 가이드라인 데이터를 수집하는 것이 좋다. 또한, A는 파팅면에 대한 캐비티와 런너의 총 투영 면적이다. 따라서 A의 값은 성형된 아이템 수와 런너의 위치에 따라 달라진다.

Example of a Calculation

Consider calculating the required mold clamping force when four molded items are obtained using PBT plastic with 30% glass fibers added.

Let us assume that the assumptions for calculation are that the pressure inside the cavity is $P = 300\ kgf/cm^2$, the projection area of one cavity is $A1 = 15.3\ cm^2$, and the projection area of the runner is $A2 = 5.5\ cm^2$.

$$F = p \times A/ 1000 = 300 \times (15.3 \times 4 + 5.5)/1000 = 20.01(tf)$$

계산의 예

4개의 성형품을 30% 유리 섬유가 첨가 된 PBT 플라스틱을 사용하여 성형할 때 요구되는 형체력을 계산해 보자.

계산을 위하여 $P = 300\ kgf/cm^2$, 하나의 캐비티의 투영 면적은 $A1 = 15.3\ cm^2$, 런너의 투영 면적은 $A2 = 5.5\ cm^2$라고 가정하자.

$$F = P \times A/1000 = 300 \times (15.3 \times 4 + 5.5)/1000 = 20.01\ (톤)$$

Therefore, an injection molding machine that has a required mold clamping force of about 20 tf is required. Giving some margin, it is considered optimum to select an injection molding machine with a 25 to 30 tf rating.

따라서, 약 20톤의 형체력을 갖는 사출성형기가 필요하다. 약간의 여유를 주면 25 ~ 30톤 급의 사출성형기를 선택하는 것이 좋다.

A 004 Mold Related Dimensions of Plastic Injection Molding Machines

A-004. 사출성형기의 금형 관련 치수

While a mold for plastic injection molding is used by installing it in an injection molding machine, at present the specifications for installing the mold in the molding machine are different for different molding machines. On the other hand, in JIS, the recommended standard specifications have been stipulated regarding the "Mold related dimensions of plastic injection molding machines" (JIS B 6701-1992). From now on, these standards are considered to be referred to for the general injection molding machines manufactured in Japan. These standards will be discussed for several courses beginning with this course.

There is only one point that has to be paid attention to and that is, in the actual design of a mold, as the final information for judgment, giving first priority to the mold installation specifications of the injection molding machine that is scheduled to be used.

Since JIS standards are only recommended standards, depending on the individual features of the injection molding machine, it is possible that the contents of the standards have been changed, and hence it is necessary to keep this point in mind.

플라스틱 사출금형은 사출성형기에 설치하여 사용되지만 현재 성형기에 금형을 설치하는 사양은 사출성형기마다 다르다. 한편, JIS에는 "플라스틱 사출성형기의 금형 관련 치수"(JIS B 6701-1992)에 관해서 권장되는 규격이 규정되어있다. 앞으로 이 표준은 일본에서 생산되는 일반적인 사출성형기에 참조사항으로 고려될 것이다. 이러한 기준들은 본 강좌를 시작으로 여러 강좌에서 논의 될 것이다.

주의할 점이 한 가지 있다. 즉, 금형의 실제 설계 시에 최종적으로 선택되어질 사출성형기의 금형 설치와 관련된 사양을 최우선적으로 고려하여야 한다.

JIS 규격은 사출성형기의 개별 특성에 따라 권장 표준 사항만으로 표준 내용이 변경되었을 가능성이 있으므로 이 점을 염두에 두어야한다.

1. Scope

These specifications stipulate the mold installation dimensions, etc. of plastic injection molding machines with a mold clamping force of 196 to 7845 kN (20 to 800 tf).

Explanation:

The names of the different parts of the injection molding machine are as shown in the figure below. (These names are merely the names used in the explanation of this standard, and do not indicate the shape or the structure of the different parts.)

In specific terms, the following items have been stipulated.

1) Spacing of the tie bar
2) Mold mounting bolt
3) Placement of the holes for mold mounting
4) Placement of the hole for the push rod (ejector rod)
5) Shape of the tip of the injection nozzle
6) Shape of the hole for the locate ring

1. 범위

이 사양은 196 ~ 7845 kN (20 ~ 800톤)의 형체력을 가진 플라스틱 사출성형기의 금형 설치 치수 등을 규정한다.

설명:

사출성형기의 각기 다른 부품들의 이름은 아래 그림과 같다. (이들 이름은 표준규격의 설명에 사용 된 이름 일 뿐이며, 다른 부품의 모양이나 구조를 나타내지는 않는다.)

구체적으로는 다음과 같은 사항이 규정되어 있다.

1) 타이바의 간격
2) 금형 체결볼트
3) 금형 체결용 구멍의 위치
4) 푸시 로드용 구멍(이젝터로드)의 위치
5) 노즐의 끝 모양
6) 로케이트 링용 구멍의 모양

A 005 Units Relating to an Injection Molding Machine

1. Injecting Device

The injecting units of a plastic injection molding machine is a unit where an injection plunger or a screw moves forward inside a cylinder to inject molten plastic into the mold.

The performance of an injecting unit is expressed by the injection rate (the volume of molten plastic that is injected in one second, cm^3/s) or the injection speed (the speed of forward movement of the plunger, ㎜/s).

In the case of super engineering plastics, there are some materials that cause loss of molded article quality unless an injecting device with a high injection rate is used, or on the other hand, there are also plastics for which the suitable injecting devices are those that can inject at a low speed in a stable manner.

It can be said that the type of injecting unit used is closely related to the molding material to be used.

The tip of an injecting unit is provided with a nozzle, and normally the tip of the nozzle is machined to a spherical shape so that it can contact closely with the spherical seat of the sprue bush of the mold, and considerations have been given so that there is no leaking of the plastic during injection. There are many types of nozzles, such as the open nozzle, the shut off nozzle, or the nozzle with a needle valve, etc.

A-005 사출성형기의 관련 장치

1. 사출장치

플라스틱 사출성형기의 사출장치는 사출 플런저 또는 스크류가 실린더 내에서 전방으로 이동하여 용융수지를 금형 내로 주입하는 장치이다.

주입장치의 성능은 사출률 (1초에 주입되는 용융수지의 부피, cm^3/s) 또는 사출속도 (플런저의 전진 속도, ㎜/s)로 표시된다.

슈퍼 엔지니어링 플라스틱의 경우, 사출속도가 빠른 사출장치를 사용하지 않으면 성형품질이 떨어지는 재료가 있고, 반면에 안정적인 상태에서 낮은 압력으로 적절한 사출 장치를 사용해야하는 플라스틱도 있다.

사용되는 사출장치의 종류는 사용되는 성형 재료와 밀접한 관련이 있다.

사출장치의 팁에는 노즐이 제공되며, 일반적으로 노즐의 팁은 구형으로 가공되어 금형의 스프루 부시의 구형부와 접촉 할 수 있으며, 주입 중에 플라스티이 새지 않는다. 개방형 노즐, 차단 노즐 또는 니들 밸브가 있는 노즐과 같은 많은 유형의 노즐이 있다.

2. Mold Clamping Unit

The mold clamping unit is a unit in charge of installing the mold in the injection machine, and of opening and closing the mold. It is necessary to clamp the mold under a high pressure so that the mold does not open when the molten plastic is being injected into the mold. Ordinarily, the size of an injection molding machine is expressed in terms of the clamping force. For example, "a molding machine with a mold clamping force of 160 tons", etc.

The methods of clamping a mold can be classified into the following two types.

1) Direct pressure type
2) Toggle type

The direct pressure type is one which the platen on the moving half is operated through a hydraulic cylinder for mold clamping. In this method, since it is possible to clamp the mold with a constant pressure applied over a wide area, this method is suitable for the molds of molded articles having a large projected area. In addition, the adjustment of the clamping force is easy, and there is the advantage that the work of installing the mold is not cumbersome even if the thickness of the mold changes.

The toggle type is one which uses the clamping force when a toggle link mechanism extends or contracts. This type has a high speed of opening and closing, and while it has the effect of shortening the cycle time, it takes time for set up because the work of adjusting the clamping force has to be done for each mold.

2. 금형 형체 장치

형체 장치는 사출성형기에 금형을 설치하고 금형을 개폐하는 장치이다. 용융수지가 금형 내로 주입될 때 금형이 개방되지 않도록 금형을 고압으로 눌러 고정시켜야 한다. 일반적으로 사출성형기의 크기는 형체력의 관점에서 표현된다. 예) "160 톤의 형체력을 갖는 사출성형기"

금형을 열리지 않게 조이는 방법은 다음 두 가지 유형으로 분류 할 수 있다.

1) 직압식
2) 토글식

직압식은 이동측 금형을 형체용 유압 실린더에 체결하여 작동시키는 타입이다. 이 방법은 넓은 영역에 일정한 압력을 가하여 금형을 형체할 수 있기 때문에 투영 면적이 큰 성형품의 금형에 적합하다. 또한, 형체력의 조정이 용이하고, 금형을 설치하는 작업이 금형의 두께가 변경되더라도 번거롭지 않은 장점이 있다.

토글식은 토글 링크를 확장시키거나 축소시킬 때 발생하는 형체력을 사용하는 타입이다. 이 타입은 개폐가 빠르며 사이클 타임을 단축시키는 효과가 있지만 형체력을 조정하는 작업은 각 금형마다 수행해야하기 때문에 셋업에 시간이 걸린다.

A mold protecting device is provided in recent mold clamping units. There are also units in which mold clamping is done at a low pressure up to a certain mold clamping process, and a protective device is provided that has a sensor that stops high pressure mold clamping if, by chance, any foreign matter or a runner, etc., has entered between the molds.

Normally, a mold clamping unit is also provided with a molded item ejecting unit which has a structure in which a knock-out bar moves forward from the moving half and causes the ejector plate of the mold to move forward thereby ejecting the molded article.

금형 보호 장치는 최신의 형체 장치에 제공된다. 형체 과정이 저압으로 이루어지게 하는 장치가 있고, 우연히 금형 사이에 이물질이나 런너 등이 들어간 경우, 그리고 고압의 형체력이 가해질 때에 멈추게 하는 센서가 있는 보호 장치가 적용된다.

일반적으로 형체 장치에는 이동측 금형에서 전방으로 전진하는 녹아웃 바가 있으며, 금형의 이젝트 플레이트가 전진하면서 성형품을 취출하는 장치가 포함되어 있다.

A 006 Fail Safe in Molds or Injection Molding Machines

In order to achieve fundamental safety of machines such as molds and injection molding machines, after first accepting the two facts of "1. Machines can fail" and "2. Operators can make mistakes", it is necessary during the stages of design, manufacturing, and modifications of machines in order to build a structure that ensures the safety of operators even if these occur by chance. For this purpose, although a "safety confirming system" is adopted, if the "safety confirming system" itself fails, since the safety of the operators is not acquired, and it is possible that a labor disaster occurs, it is necessary to have characteristics that even when the "safety confirming system" fails, always the equipment is on the safe side (stopping the machine in a state in which it does not cause labor disasters).

A-006 금형과 사출성형기의 안전장치

금형 및 사출성형기와 같은 기계의 근본적인 안전을 확보하기 위해 "1. 기계는 고장날 수 있음"과 "2. 작업자는 실수를 할 수 있음"의 두 가지 사실을 처음부터 인정하고, 우연히 발생하더라도 작업자의 안전을 보장하는 구조를 만들기 위해서 기계의 설계, 제조 및 개조 단계에서부터 고려할 필요가 있다. 이를 위해 "안전 확인 시스템"을 채택하고 있지만 "안전 확인 시스템" 자체가 고장 나면 작업자의 안전을 확보하지 못하고, 노동 재해가 발생할 가능성이 있다. 따라서 "안전 확인 시스템"이 고장난 상황에서도 그 시스템은 항상 안전한 상태가 되어야 한다.(노동 재해가 발생하지 않는 상태에서 기계를 정지시킴).

In order to achieve fail safe, it is necessary to carry out the design, manufacturing, and modifications of molds, injection molding machines, automatic machines, etc., while following the "fundamental rules for achieving fail safe" listed below.

안전장치를 확보하기 위해서 아래에 기재된 「안전장치 확보를 위한 기본 규칙」에 따라 금형, 사출성형기, 자동 기계 등의 설계, 제조, 개조를 실시해야 한다.

1) As a rule, the following are the control mechanisms that are the target of implementing fail safe.
   a. Restart preventing circuit.
   b. Interlock circuits for guarding
   c. Circuits for sudden stop
   d. Circuits for emergency stop
   e. Circuits for preventing exceeding the limits
   f. Circuits for monitoring operations
   g. Circuits for hold stop monitoring
   h. Circuits for speed monitoring
   i. Hold to run circuits

1) 일반적으로 안전장치 실행을 위한 기계의 제어 메커니즘은 다음과 같다.
   a. 재시작 방지 회로
   b. 감시를 위한 인터록 회로
   c. 급정지 회로
   d. 비상정지 회로
   e. 제한 초과 방지 회로
   f. 운영 모니터링 회로
   g. 정지 감시용 회로
   h. 속도 모니터링 회로
   i. 회로 실행 보류

2) As a rule, the control mechanism should be designed to have "asymmetrical error characteristics". Asymmetrical error characteristics are the characteristics by which even when a system or a component constituting the system fails, the frequency of failures on the safe side is far greater than the frequency of failures on the danger side, or the characteristics of failing only on the safe side.

2) 일반적으로 제어 메커니즘은 "비대칭 오류 특성"을 갖도록 설계해야한다. 비대칭 오류 특성은 시스템 또는 시스템을 구성하는 구성 요소가 고장 나더라도 위험부위의 파손보다는 안전부위의 파손이 훨씬 잘 일어나거나, 또는 안전부위에서만 파손이 주로 일어나도록 하는 특성이다.

3) When adopting electronic control equipment such programmable controllers for the control mechanism, use only those having "asymmetrical error characteristics".

3) 제어 메카니즘을 위하여 프로그램 작동이 가능한 제어기와 같이 전자 제어 기구를 채택할 경우, "비대칭 오류 특성"이 있는 것을 사용하자.

4) Danger or fault conditions are not mistakenly reported as safe by making safety information correspond to a high energy state, danger signals and fault signals correspond to a low energy state.

4) 안전 정보가 고 에너지 상태, 위험 신호 및 오류 신호가 저에너지 상태에 해당하도록 함으로써 위험 또는 오류 조건을 안전하다고 잘못보고하지 않는다.

5) Safety information should be conveyed in an information conveying mode so that, unless safety information is input to the system, an operation permitting signal is not issued by mistake.

5) 안전 정보는 정보 전달 모드로 전달되어야하므로 안전 정보가 시스템에 입력되지 않으면 실수로 작동 허용 신호가 발생되지 않는다.

6) In order to acquire resistance to the maximum environmental noise that can be predicted, the safety information should be made to possess sufficient energy.

6) 예측 가능한 최대 환경 소음에 대한 내성을 확보하기 위해서 충분한 에너지를 보유하도록 안전 정보를 만들어야한다.

A fail safe design philosophy is required during the design of molds that are opened and closed at a high speed during high cycle rates or of molds of large sizes.

고 사이클 률에서 매우 빠르게 열고 닫히는 금형을 설계하거나 대형 금형을 설계할 때에는 오류 방지 설계 철학이 필요하다.

A 007 Power Source and Controlling Method of an Injection Molding Machine

A-007 사출성형기의 동력원 및 제어 방법

1. Power Source of an Injection Molding Machine

The following three types are the most frequently used power sources of injection molding machines.

1) Hydraulic type
2) Electric motor driven type
3) Hybrid type

1. 사출성형기의 동력원

다음 세 가지 유형이 사출성형기의 가장 자주 사용되는 동력원이다.

1) 유압식
2) 전기 모터 구동식
3) 하이브리드식

The hydraulic type is the one that has been used most often so far ever since the injection molding machine was developed. The mold is opened and closed by the hydraulic pressure of the working fluid supplied by driving a hydraulic pump by an electric motor, thereby operating the mechanism during injection. While this method has advantages such as it is possible to achieve high mold clamping forces relatively easily, it also has drawbacks such as it is difficult to carry out precision control, it is difficult to operate several actuators in parallel at the same time, etc.

The electric motor driven type uses a method of driving using an AC servomotor under computer numerical control (CNC). In this method, it is possible to realize accurate operation with no use of operating oil, and also, it is possible to carry out clean and silent operation. There is also the advantage over the hydraulic type that the running cost is low because the electric power consumption is about 40 to 50% of the hydraulic type. Recently, electric motor driven types with mold clamping forces up to 2000 tons haven been developed, and this method will become the major one from now on at mold clamping forces of 350 tons or less.

The hybrid type is one in which both the hydraulic method and the electric motor driven method driving the machine are used, with the intention of obtaining the advantages of both methods. The general pattern is to make either the injection side or the movable side hydraulic type.

유압식은 사출성형기가 개발된 이래로 지금까지 가장 자주 사용되는 타입이다. 금형은 전기 모터에 의해 유압 펌프를 구동함으로써 공급되는 작동 유체의 유압에 의해 개폐되어, 사출 중에 기구를 작동시킨다. 이 방법은 비교적 용이하게 높은 형체력을 얻을 수 있는 이점을 가지지만 정밀 제어가 곤란하고, 복수의 액추에이터를 동시에 병렬로 동작시키는 등의 어려움이 있다.

전기 모터 구동 방식은 컴퓨터 수치 제어 (CNC) 방식의 AC 서보 모터를 사용하여 구동하는 방식을 사용한다. 이 방법은 작동유를 사용하지 않고 정확한 작동을 실현할 수 있을 뿐만 아니라 청결하고 조용한 작동을 수행 할 수 있다. 전력 소비가 유압식의 약 40 ~ 50%이기 때문에 운전비용이 저렴하다는 이점이 있다. 최근에 2000 톤까지의 형체력을 가진 전기 모터 구동식 사출성형기가 개발되었으며, 이 방식은 305 톤 이하의 형체력에서 주로 사용된다.

하이브리드 타입은 유압식 타입과 전기 모터 구동식 타입의 장점을 접목시킨 사출성형기이다. 일반적인 패턴은 사출부분 또는 가동부분 중의 하나를 유압식으로 만드는 것이다.

4) Electric Control Device

A control computer is used for carrying out process control for controlling the operation of the injection molding machine and the molding conditions, and for

4) 전기 제어 장치

제어 컴퓨터는 사출성형기의 작동 및 성형 조건을 제어하고, 센서 검출을 이용하여 피드백 제어를 수

carrying out feedback control using sensor detection, and a control device is used for controlling the screen display, etc. The molding conditions, etc. can be recorded in a computer, and it is also possible to record them in an external computer connected via a LAN, or in an external memory.

행하며, 제어 장치가 스크린 디스플레와 공정 제어를 수행하는데 사용된다. 성형 조건 등을 컴퓨터에 기록 할 수 있고, 또한 LAN을 통해 연결된 외부 컴퓨터 또는 외부 메모리에 기록 할 수도 있다.

A 008 Features of Motor Driven Injection Molding Machines

A-008 모터 구동 사출성형기의 특징

Recently, the use of fully electric motor driven injection molding machines is spreading rapidly particularly in the case of smaller sized machines. Research on the realization of fully electric motor driven injection molding machines started about 15 years ago, the great advancement in the their functions in the past few years is considered to be the driving force for their coming into increasingly wider use.

The following are the features of a fully electric motor driven injection molding machine.

최근에 소형의 사출성형기에 전기 모터를 사용하는 사출성형기의 사용이 급속히 확산되고 있다. 15년 전부터 전기 모터 구동 사출성형기의 사용에 관한 연구가 시작된 이래로, 지난 몇 년 동안 그들의 기능이 크게 발전한 것은 점점 더 폭넓게 사용될 수 있었던 원동력이었다.

다음은 전기 모터 구동 사출성형기의 기능이다.

1) The electric energy consumption very much smaller compared to a hydraulic type molding machine. A reduction of 40 to 50% is possible. That is, there is the effect of greatly reducing the monthly electricity usage charges.

1) 전기 에너지 소비량은 유압식 성형기에 비해 매우 적다. 40 ~ 50%의 감소가 가능하다 즉, 월별 전기 사용료를 크게 줄이는 효과가 있다.

2) Since the control of the injection molding machine can be made accurately using servo motors, it is possible to achieve stabilization of the molding conditions.

2) 서보 모터를 사용하여 사출성형기의 제어를 정확하게 할 수 있기 때문에 성형 조건의 안정화를 도모 할 수 있다.

3) Injection molding operations can be made a low noise generation level.

4) Since normally the operations of mold opening and closing, injection, ejection, and weighing can be done using four independent motors, it becomes possible to make the molding operations progress in parallel, and as a result, it is possible to shorten the molding cycle.

5) The operations of accurate control of the amount of ejection, multi-stage ejection, etc. are possible that were difficult in the case of hydraulic molding machines.

6) While it was difficult to increase the injection speed, recently, it has become possible to meet conditions of a level from 1000 ㎜/s to 2000 ㎜/s.

7) Since no oil is used, the molding environment becomes clean, and this is very convenient for molding food containers or products for medical use.

8) At present the price is higher than those of hydraulic molding machines.

3) 사출성형 작업은 낮은 소음이 발생 한다.

4) 일반적으로 4개의 독립적인 모터를 사용하여 금형 개폐, 사출, 취출 및 계량 작업을 수행 할 수 있기 때문에 성형 작업이 병렬로 진행될 수 있고, 결과적으로 성형 사이클 타임을 단축 할 수 있다.

5) 유압 성형기의 경우에는 어려웠던 취출량, 다단 취출 등의 정확한 제어가 가능하다.

6) 사출속도를 높이기는 어려웠으나 최근에는 1000 ㎜/s에서 2000 ㎜/s까지의 조건을 만족 할 수 있게 되었다.

7) 기름을 사용하지 않기 때문에 성형 환경이 깨끗해져 의료용 제품, 식품용기 성형품 성형에 매우 편리하다.

8) 현재 유압 성형기보다 가격이 비싸다.

Although the use of fully electric motor driven injection molding machines has progressed more in Europe, United States, and Canada than in Japan, it is expected that the advantages of the electric type machine will be exploited more in the case of the injection molded products that are used in new products in IT related industries, and automobile industries.

As the injection molding machines become advanced, even increases in the technological level of molds will gradually be expected. For example, it may be necessary to enhance the level of precise positioning mechanisms, or air vent mechanisms, mold temperature control functions, etc. Even in the world of standard components for molds, it is expected that new products will be developed meeting such needs.

전기 모터 구동 사출성형기의 사용이 일본보다 유럽, 미국 및 캐나다에서 더 많이 진행되었지만, IT 관련 산업 및 자동차 산업의 신제품에 사용되는 사출성형 제품의 경우 전기 모터 구동 사출성형기의 장점이 더 많이 활용 될 것으로 기대된다.

사출성형기가 발전함에 따라 금형 기술 수준의 향상도 점차 기대된다. 예를 들어, 정밀한 포지셔닝 메커니즘이나 에어벤트 메커니즘, 금형 온도 제어 기능 등을 향상시킬 필요가 있을 수 있다. 금형 표준 부품 분야에서도 이러한 요구를 충족하는 신제품이 개발될 것으로 예상된다.

A 009 Mold Design for Electric Injection Molding Machines

Electric injection molding machines are coming into wide use in the case of injection molding machines with mold clamping forces of 1800 t or less. In order to bring out to the maximum extent the excellent features possessed by fully electric injection molding machines, it is very important to carry out mold design giving attention to the following factors.

1. Accurate positioning

In a fully electric driven injection molding machine, although it is possible to carry out at high speeds the opening and closing of molds, in order to prevent breakage or abnormal wear at the time of contact between core pins while carrying out accurate positioning of the cavities on the movable side and on the fixed side, it is recommended to install precision positioning guides.

The recommended Misumi standard components are:

- Tapered tweezers TPN, etc.
- Tapered block sets TBS, etc.
- Side straight block set TSSB, etc.

A-009 전동 사출성형기용 금형 설계

1800 톤 이하의 형체력을 가진 사출성형기의 경우 전동 사출성형기가 널리 보급되고 있다. 완전 전동식 사출성형기가 갖고 있는 뛰어난 기능을 최대한으로 발휘하기 위해서 다음과 같은 점에 주의를 기울여 금형설계를 하는 것이 매우 중요하다.

1. 정확한 위치

완전 전동 사출성형기에서 금형을 빠르게 열고 닫는 것이 가능할지라도, 이동측 금형과 고정측 금형의 캐비티들이 정확한 위치 결정을 진행되는 동안 코어 핀들 사이의 접촉시 파손이나 비정상석 마모를 방지하기 위해서 정밀 위치 결정 가이드를 설치하는 것이 좋다.

권장 미스미 표준 구성 요소는 다음과 같다.

- TPN 등 Tapered 핀셋
- Tapered 블록은 TBS 등을 설정
- 측 직선 블록은 TSSB 등을 설정

2. Smooth operation of the ejector plate

In a fully electric motor driven injection molding machine, although it is possible to carry out operations of ejector plates such as high speed operation and multiple stage ejection, if the construction is not designed so that the operation of the ejector plate is smooth, biting of the ejector pin or biting of the return pin are likely to occur.

Therefore, it is recommended to use an ejector guide system.

The following Misumi standard components are being used widely:

- Precision ejector guide pin EGH, etc.
- Ejector guide bush EGBH (Plain type)

Further, it is recommended to install a switch for confirming the returning of the ejector plate.

2. 이젝터 플레이트의 부드러운 작동

전동 모터 구동의 사출성형기에는 고속 작동 및 다단 취출과 같은 이젝터 플레이트의 작동을 수행 할 수 있지만, 이젝터 플레이트의 작동이 원활하도록 설계되지 않으면 이젝터 핀 또는 리턴 핀의 손상이 발생할 수 있다.

따라서 이젝터 가이드 시스템을 사용하는 것이 좋다.

다음 Misumi 표준 컴포넌트가 널리 사용되고 있다 :

- 정밀 이젝터 가이드 핀 EG 등
- 이젝터 가이드 부시 EGBH (일반 타입)

또한 이젝터 플레이트의 복귀 확인 스위치를 설치하는 것이 좋다.

3. Smooth operation of the slide core

Smooth operation of the slide core is also important. In order to prevent biting, it is recommended to provide a lubricating groove in the guide rail, or to use a non-lubricating type material (oil free metal).

The following Misumi standard components are recommended:

- GR5SCM (Guide rail with groove)
- SRTL (Non-lubricating guide rail)
- CGZ (Non-lubricating center rail)
- STW (Non-lubricating slide plate).

3. 슬라이드 코어의 부드러운 작동

슬라이드 코어의 원활한 작동 또한 중요하다. 손상을 방지하기 위해 가이드 레일에 윤활 홈을 붙이거나 비 윤활 유형 재료 (오일이 없는 금속)를 사용하는 것이 좋다.

다음과 같은 Misumi 표준 구성 요소를 권장한다.

- GR5SCM (가이드 홈에 레일)
- SRTL(가이드 레일을 비는 윤활)
- CGZ(비 윤활 센터 레일)
- STW(슬라이드 플레이트를 비는 윤활).

4. Stabilization of the cavity surface temperature

In order to stabilize the molding conditions, it is very important to stabilize the mold cavity surface temperature. In order to achieve this, it is recommended to use the following standard components:

- HIP (Thermal insulation plate), etc.
- M-HTM3021 (For cartridge heaters)
- Mold temperature controller: PID control type

4. 캐비티 표면 온도의 안정화

성형 조건을 안정화시키기 위해서 금형 캐비티 표면 온도를 안정화하는 것이 매우 중요하다. 이를 위해 다음과 같은 표준 구성 요소를 사용하는 것이 좋다.

- HIP (보온 판) 등
- M-HTM3021(카트리지 히터 용)
- Mold 온도 컨트롤러 : PID 제어 방식

B. 사출성형조건

CHAPTER B

B 001 What is the Molding Shrinkage Phenomenon?

In the injection molding of thermoplastic plastics, it is possible to obtain a molded product with the desired dimensions using the mold shrinkage phenomenon. Mold shrinkage is the phenomenon where the volume of the molten plastic filled inside the cavity of a mold is shrinking at the time as being cooled and solidifying.

The extent of this shrinkage is called the "molding shrinkage factor", and if this molding shrinkage factor is known accurately both scientifically and by experience, by preparing the mold making the dimensions of the cavity a little larger by the amount of shrinkage, it is possible to form the molded item by so that it has the intended dimensions.

The value of the molding shrinkage factor is generally a number in the range of about 2/1000 to 20/1000 (about 0.2 to 2%).

If the molding shrinkage factor is expressed by the symbol α (alpha), it can be defined by the following equation 1.

$$\alpha = (L_0 - L)/L_0 \cdots \quad \text{(Eq.1)}$$

Where, L_0: the cavity dimensions (㎜) L: Dimensions (in ㎜) of the molded product at room temperature (usually 20℃).

Further the molding shrinkage factor is affected by the following factors.

B-001 성형 수축 현상이란 무엇인가?

열가소성 플라스틱의 사출성형에서 성형수축 현상을 이용하여 원하는 크기로 성형된 제품을 얻을 수 있다. 성형 수축은 금형의 캐비티 내부에 채워진 용융수지가 점차 냉각되고, 고화되면서 수축하는 현상이다.

이 수축의 정도를 "성형 수축률" 이라고 하며, 이 성형 수축률이 과학적이며, 경험적으로 정확하게 알려져 있다면, 수축되는 양만큼 큰 캐비티를 준비하여 성형한다면 의도하는 크기를 갖춘 성형품을 성형할 수 있을 것이다.

성형 수축률은 일반적으로 2/1000 ~ 20/1000 (약 0.2 ~ 2%) 범위이다.

성형 수축률을 α(알파)라 하면, 수축률은 식 1과 같이 정의 할 수 있다.

$$\alpha = (L_0 - L)/L_0 \cdots \quad \text{(식 1)}$$

여기서, L_0은 캐비티의 원래 치수(㎜), L은 상온(일반적으로 20℃)에서의 성형품 치수(㎜)를 말한다.

또한 성형 수축률은 다음과 같은 요인에 의해 영향을 받는다.

(1) Type of molding material

The range of the basic shrinkage factor is determined by the type of plastic material being used. However, there will be fine differences depending on the material manufacturer and the grade of the material.

(1) 수지의 종류

기본 수축률의 범위는 사용되는 플라스틱 재료의 종류에 의해 결정된다. 그러나 플라스틱의 제조업체 및 수지의 그레이드에 따라 미소한 차이가 있을 수 있다.

(2) Cavity surface temperature

The molding shrinkage factor varies depending on the cavity surface temperature during injection molding. In general, the shrinkage factor tends to be large when the temperature is high.

(2) 캐비티 표면 온도성형

수축률은 사출성형 할 때 캐비티 표면 온도에 따라 달라진다. 일반적으로 수축률은 캐비티 표면 온도가 높을수록 커지는 경향이 있다.

(3) Maintained pressure × pressure maintenance time

The molding shrinkage factor varies depending on the magnitude of the pressure maintained after plastic injection and the time of maintaining that pressure. In general, there is trend in the shrinkage factor becoming smaller when the maintained pressure is high and the pressure maintenance time is long.

(3) 보압 × 보압 유지시간

성형 수축률은 수지 충전 후 보압의 크기와 보압이 유지되는 시간에 따라 변한다. 일반적으로 보압이 높고, 보압 유지시간이 길수록 수축률이 작아지는 경향을 갖고 있다.

(4) Wall thickness of the molded item

The shrinkage factor also varies depending on the wall thickness of the molded item. There is a trend in the shrinkage becoming larger as the wall thickness becomes larger.

(4) 성형품의 살 두께

수축은 성형품의 살 두께에 따라 변한다. 살 두께가 두꺼워 질수록 성형 수축은 더 커지는 경향을 갖는다.

(5) Gate shape

The shrinkage factor varies depending on the gate shape and the gate size. In general, there is a trend in the shrinkage becoming smaller as the cross-sectional area of the gate becomes larger.

(5) 게이트 형상

성형 수축은 게이트의 형상과 크기에 따라 달라진다. 일반적으로 게이트의 단면적이 작아질수록 수축이 커지는 경향이 있다. 핀 포인트 게

There is also a trend in the shrinkage becoming smaller in the case of a side gate rather than in the case of a pinpoint gate or a submarine gate.

이트나 서브마린 게이트보다는 사이드 게이트의 경우가 수축이 더 작아지는 경향이 있다.

(6) Presence or absence of additive materials to the molding material

It is very common that there is a large difference in the shrinkage factor between natural materials and materials having glass fibers.

There is a trend in the shrinkage factor being smaller in the case of materials with glass fibers. In actuality, the molding shrinkage factor for mold design is determined by comprehensively investigating the above conditions.

(6) 성형 재료 내의 첨가재의 유무

일반적으로 순수 수지와 유리 섬유가 첨가된 수지 사이의 성형 수축은 큰 차이가 있다. 유리 섬유가 첨가된 수지의 경우 성형 수축이 더 작아지는 경향이 있다. 실제로 금형설계를 위한 성형 수축은 포괄적으로 위의 조건을 조사하여 결정된다.

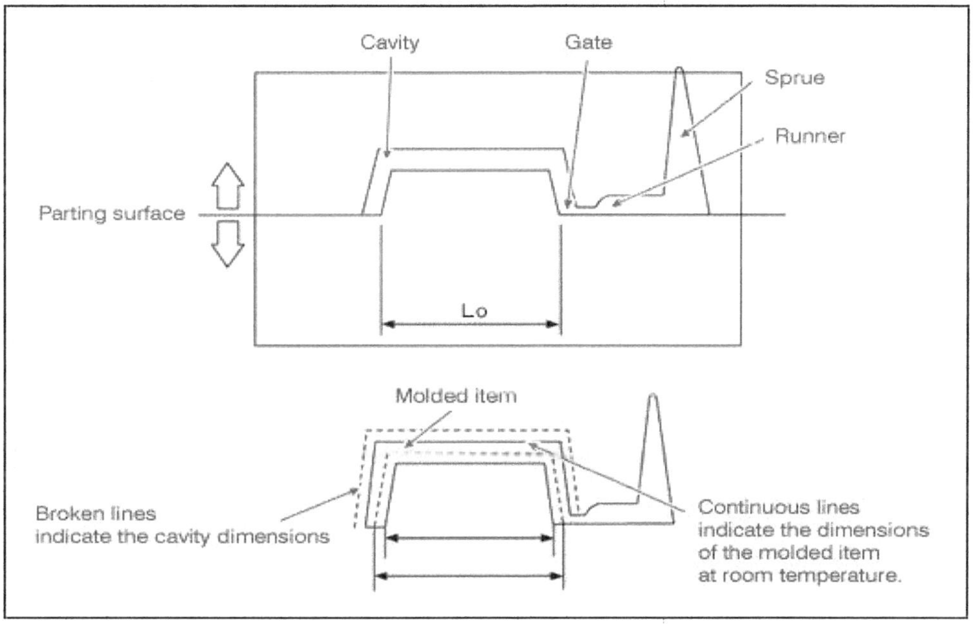

B 002 Basic Theory of Molding Shrinkage Ratio

Molding shrinkage ratio is one of the most important factors in the design of molds for plastic injection molding.

The molding shrinkage ratio α can be expressed by the following equation.

$$\alpha = (L_0 - L)/L_0, \text{ where,}$$

α : Molding shrinkage ratio (no units)

L_0 : Mold dimensions (㎜)

L : Product dimensions (㎜)

However, in actuality, since the volume shrinks when the molten plastic cools, if the concept of volume shrinkage ratio is used, it can be expressed by the following equation.

$$\alpha v = (V_0 - V_1)/V_0, \text{ where}$$

αv : Volume shrinkage ratio (no units)

V_0 : Volume of the mold cavity (㎣)

V_1 : Volume of the molded product (㎣)

Here, if the volumes V_0 and V_1 are considered as three dimensional objects, it is possible to express them using the following equations.

$$L_0 = \sqrt[3]{V_0}$$

$$L_1 = \sqrt[3]{V_1}$$

B-002 성형 수축률의 기본 이론

성형 수축률은 플라스틱 사출 금형설계에서 가장 중요한 요소 중 하나이다.

성형 수축률 α는 다음 식으로 나타낼 수 있다.

$$\alpha = (L_0 - L)/L_0, \text{ 여기서,}$$

α : 성형 수축률 (단위 없음)

L_0: 금형 크기 (㎜)

L : 성형품 크기 (㎜)

그러나 실제로 용융수지가 냉각될 때 체적이 줄어들기 때문에 체적 수축률의 개념을 사용하면 다음 식으로 나타낼 수 있다.

$$\alpha V = (V_1 - V_0)/V_0 \text{ 여기서,}$$

αv : 체적 수축률 (단위 없음)

V_0 : 금형 캐비티의 체적 (㎣)

V_1 : 성형품의 체적 (㎣)

여기서, 체적 V_0 및 V_1은 삼차원 물체로 가정하는 경우 여기서, 다음의 방정식을 사용하여 표현할 수 있다.

$$L_0 = \sqrt[3]{V_0}$$

$$L_1 = \sqrt[3]{V_1}$$

Therefore,

$$\alpha v = (V_0 - V_1)/V_0$$
$$= 1 - (V_1/V_0)\ 1/3$$
$$= 1 - (v_1/v_0)\ 1/3,\ \text{where,}$$

v_1 : Specific volume at room temperature (㎣/g)

v_0 : Specific volume under molding conditions (㎣/g)

In addition, it is known that the state equation of the molten plastic can be expressed as follows.

(P+πi) (v-ω) =R'T, where,

v : Specific volume of the molten plastic (㎣/g)

πi : Internal pressure (atm)

ω : Specific volume at absolute zero degrees (㎣/g)

R' : R/M (Corrected gas constant atm ㎣/g • mol)

M : Molecular weight in units of molecular activity

If αv =1-(v1/v0)1/3 is substituted in the above equation, we get:

α=1-3√y

= (1-y)/3+ (1-y) 2/9

y=v0 (P+πi) /R'T+ω (P+πi)

From this, it is also clear theoretically that the Pressure P and the temperature T of the molten plastic have a large influence on the molding shrinkage ratio α.

따라서,

$$\alpha V = (V_0 - V_1)/V_0$$
$$= 1 - (V_1/V_0)\ 1/3$$
$$= 1 - (v_1/v_0)\ 1/3,\quad \text{여기서}$$

V_1 : 실온 (㎣/g)에서의 비체적

V_0 : 성형 조건(㎣/g)에서의 비체적

또한, 용융수지의 상태방정식은 다음과 같이 표현 될 수 있다.

(P+πi)(v-ω)=R'T, 여기서,

v : 용융수지의 비체(㎣/g)

πi : 내부 압력

ω : 절대 0도 에서의 비체적(㎣/g)

R' : R/M(보정 기체 상수, ㎣/g • mol)

M : 분자량

위의 방정식에 αv =1-(v1/v0)1/3 을 대입하면 다음과 같다.

α=1-3√y

=(1-y)/3+(1-y)2/9

y=v0(P+πi)/R'T+ω(P+πi)

이로부터 이론적으로 용융수지의 압력 P 및 온도 T가 성형 수축률 α에 큰 영향을 미친다는 것을 알 수 있다.

* Reference: "Injection molding technology exploiting the material characteristics" (in Japanese, by Seiichi Honma, Sigma Publications).

B 003 What is an Injection Molding Cycle?

"Molding cycle" is an index for evaluating the production capacity of injection molding. A molding cycle is the unit of time from starting one injection molding operation until it is ended.

In more concrete terms, a molding cycle consists of the following.

Molding cycle (seconds / shot)

= Mold closing time + Filling time + Holding(packing) time +Cooling time (including the charging time) +Mold opening time + Product taking out time

As the molding cycle becomes shorter, it becomes possible to increase the quantity of production of the molded product per unit time.

Example :

Consider that there is a mold whose molding cycle at present is 10 seconds. By modifying the cooling circuit of the mold, the cooling efficiency was increased, the cooling time was reduced, and as a result, it was possible to make the molding cycle 9 seconds.

What kinds of factors are involved in this case?

B-003 사출성형 사이클이란?

"성형 사이클"은 사출성형의 생산 능력을 평가하는 지표이며, 성형 사이클은 사출성형 1 회 작업을 시작한 후 종료 될 때까지 걸리는 시간 단위이다.

보다 구체적으로 성형 사이클은 다음과 같이 구성된다.

성형 사이클(초/회)

= 금형 닫힘시간 + 충전시간 + 보압시간 + 냉각시간(용융수지 충전시간 포함) + 금형 형개시간 + 취출시간

성형 사이클이 짧을수록 단위시간당 성형품의 생산량을 증가시킬 수 있다.

예:

현재 성형 사이클이 10 초인 금형이 있다고 가정하자. 금형의 냉각회로를 변경하여 냉각 효율이 향상되었고, 냉각시간이 단축되었다. 결과적으로 성형 사이클을 9 초로 할 수 있었다.

이 경우 어떤 종류의 요인이 관련되는가?

Example Answer:

Since the molding cycle was reduced from 10 seconds to 9 seconds, it became shorter by 1 second.

Although it is merely 1 second, if we think of it as a rate of change, it becomes - Rate of change = (10−9) seconds / 10 seconds ×100 = 10%

And hence the productivity is improved by 10%.

If a large number of pieces are obtained from a mold such as 12 or 24 pieces per shot, and if there are many similar molds, the productivity of the entire factory will be improved by 10%, and when converted into money in terms of the production cost, it can become a very large amount.

In other words, evaluating the molding cycle directly results in strengthening the profitability of the factory, and not merely reducing by 1 second but leading to an extremely large good effect.

Therefore, in the design of a mold, it is very important to consider at all times by how much the molding cycle can be shortened.

In addition, it is also very important that the mold designer estimates the target molding cycle.

An excellent mold designer always estimates the target molding cycle.

답 :

성형 사이클이 10 초에서 9 초로 단축되었으므로 성형 사이클이 1 초 단축되었다.

단지 1 초일지라도 변화율로 생각하면 변화율 = (10-9) 초 / 10 초 × 100 = 10% 향상된다.

한번 성형으로 12개나 24개의 많은 성형품이 생산되고, 비슷한 금형이 많이 존재한다면 공장 전체의 생산성은 10% 향상되고, 생산비용을 돈으로 환산하면 매우 큰 금액이다.

즉, 성형 사이클을 직접 평가하면 공장의 수익성이 강화되고, 1 초만 줄이는 것이 아니라 매우 큰 효과로 나타난다.

따라서 금형설계 시 성형 사이클이 얼마나 단축될 수 있는지 항상 고려해야한다.

또한 금형설계자가 목표 성형 사이클을 예측하는 것은 매우 중요하다.

우수한 금형설계자는 항상 목표 성형 사이클을 예측한다.

B 004 Constituent factors of an Injection Molding Cycle

In general, the following are the constituent factors of an injection molding cycle.

1. Mold closing time

This is the time until the mold is closed. The times of opening and closing the mold vary depending on the speeds of the mold opening stroke and the mold closing stroke of the injection molding machine.

Since the mold has considerable mass, if the mold closing speed is made too fast recklessly, it is possible that the mold may break at the time of closing the mold due to the kinetic energy, and hence it is necessary to consider applying brakes.

2. Filling time

The filling time is the time taken until the molten plastic flowing in from the sprue completely fills the inside of the cavity.

The filling time is determined by the injection speed of the injection molding machine (actually, by the injection volume per unit time = injection rate (cm^3/sec) which is determined by the movement speed of the injection screw or piston and the diameter of the injection cylinder).

In addition, this is also affected by the viscosity of the molten plastic.

B-004 사출성형 사이클의 구성 요소

일반적으로 다음은 사출성형 사이클의 구성 요소이다.

1. 금형 닫힘시간

이것은 금형이 닫힐 때까지의 시간이다. 금형의 열림과 닫힘시간은 사출성형기의 금형 열림 스트로크 및 금형 닫힘 스트로크의 속도에 따라 다르다.

금형이 상당히 무거울 경우가 있기 때문에 금형 닫힘 속도가 너무 빠르면 운동 에너지로 인해 금형을 닫을 때 금형이 파손될 가능성이 있으므로 브레이크의 적용을 고려해야한다.

2. 충전시간

충전시간은 스프루에 유입된 용융수지가 캐비티의 내부를 완전히 채울 때까지 걸리는 시간이다.

충전시간은 사출성형기의 사출속도(실제로, 단위 시간당 사출 체적은 사출률(cm^3/sec)로서 사출성형기의 스크류 이동속도 또는 사출성형기의 실린더 직경에 의해서 정의된다.)에 의해서 정의 된다.

또한, 용융수지의 점도에 영향을 받는다.

3. Holding time (Packing time)

The holding time is the duration of time during which pressure is being applied after filling the inside of the cavity with molten plastic until the gate part solidifies.

3. 보압시간 (팩킹시간)

보압시간은 용융수지로 캐비티 내부를 충전 한 후 게이트 부분이 고화 될 때까지 압력이 가해지는 시간이다.

4. Cooling time

The cooling time is the time duration after holding has been completed while the cooling takes place until the molded product, the sprue, and the runners solidify to some extent.

In general, the cooling time tends to be the longest time in a molding cycle.

4. 냉각시간

냉각시간은 보압이 완료 된 후 성형품, 스프루, 런너가 고화 될 때까지 지속되는 시간이다.

일반적으로 냉각시간은 성형 사이클에서 가장 긴 시간을 차지한다.

5. Charging time

The charging time is the time required for measuring the amount of molten plastic for the next injection molding. In general, the charging is carried out simultaneously with the cooling and the charging time is shorter than the cooling time.

5. 계량시간

계량시간은 다음 사출성형을 위하여 용융수지를 계량하는데 필요한 시간이다. 일반적으로 용융수지의 계량은 냉각과 동시에 수행되고, 계량시간은 냉각시간보다 짧다.

6. Mold opening time

This is the time duration until the mold is opened completely. The mold opening and closing times vary depending on the speeds of the mold opening stroke and the mold closing stroke of the injection molding machine.

6. 금형 열림시간

이 시간은 금형이 완전히 열릴 때까지의 시간이다. 금형 열림시간은 사출성형기의 금형 열림 스트로크와 금형 닫힘 스트로크의 속도에 따라 달라진다.

7. Taking out time

This is the time taken for taking out the molded product from the cavity. This includes the ejection time and the time of operation of the take-out robot.

7. 취출시간

이 시간은 캐비티로부터 성형품을 취출하는 데 걸리는 시간이다. 즉, 취출시간과 취출 로봇의 동작시간을 포함한다.

B 005 Stabilizing the Clamping Force of Single Cavity Mold

In a single cavity mold, very often the cavity is placed only on one half of the mold plate with respect to the sprue. (See the example in Fig. 1.)

If the size of the mold plate becomes large, since the pressure receiving plate is only on one side, the balance of the mold clamping force becomes disturbed at the time of injection molding, and burrs may appear on the periphery of the molded product.

B-005 단일 캐비티 금형의 형체력 안정화

단일 캐비티 금형에서 캐비티는 거의 스프루에 대하여 형판의 한쪽 절반에만 위치한다.(그림 1의 예 참조)

형판의 크기가 커지면 압력이 한쪽에만 가해지기 때문에 형체력의 균형이 트러져 버리거나 성형품의 주변부에 버가 생길 수 있다.

Fig. 1

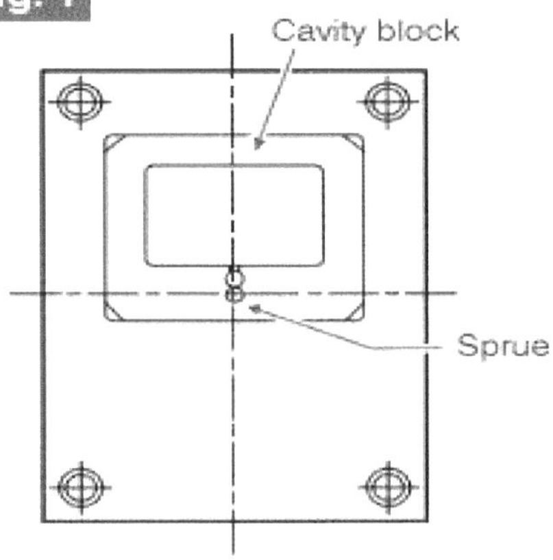

Fig. 2

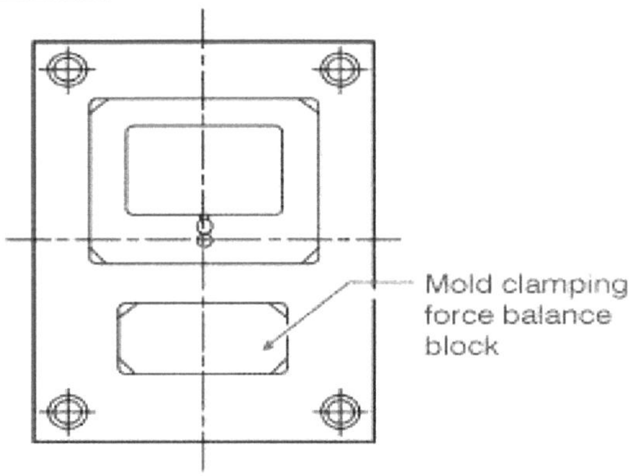

If the clamping force is set unduly high in order to prevent the generation of burrs in such situations, or if the filling pressure is set lower, it is likely that a different problem occurs in some unexpected part.

When it is judged that the balance of the mold clamping force is disturbed in a single cavity mold, an appropriate countermeasure is to place a balance block on the symmetrically opposite side with respect to the sprue as is shown in Fig. 2.
Usually, since the parting surface of the cavity and core is projecting by about 5 to 20 micrometers from the parting surface of the mold plate, by placing a balance block, it is possible to put the sealing of the cavity and core parting surface in the desirable condition.

The shape of the balance block can be rectangular or can be circular, and as larger a surface area of receiving the mold clamping force as possible will make the balance more stable.

이러한 상황에서 버가 발생하지 않도록 형체력을 너무 높게 설정하거나 충전압력을 낮게 설정하면 예기치 않은 부분에서 다른 문제가 발생할 가능성이 있다.

단일 캐비티 금형에서 형체력의 균형이 흐트러진 것으로 판단되면, 적절한 대책은 그림 3과 같이 스프루에 대해 반대쪽에 균형 블록을 배치하는 것이다.

일반적으로 캐비티와 코어의 파팅면은 형판의 파팅면으로 부터 약 5 ~ 20 마이크로미터 돌출되어 있기 때문에 밸런스 블록을 배치함으로써 캐비티와 코어의 파팅면 실링을 원하는 상태로 할 수 있다.

밸런스 블록의 형상은 직사각형 또는 원형 일 수 있으며, 가능한 형체력을 받는 표면적이 클수록 균형이 보다 안정적이다.

B 006 The Flow Ratio (L/t) of Plastic

B-006 플라스틱의 유동 비율 (L / T)

In order to fill the plastic inside the cavity, it is necessary to push it inside the cavity while applying pressure from the injection cylinder. When the plastic is in the heated and molten state, it is a fluid with some viscosity. However, the viscosity starts increasing as the plastic reaches the cavity while flowing through the sprue and the runner because it loses heat to the surface of the mold. If the viscosity increases beyond a certain limit, the leading part solidifies due to cooling,

캐비티 내부를 플라스틱으로 채우기 위해서 사출 실린더에서 압력을 가하며 용융수지를 캐비티 내부로 사출한다. 플라스틱이 가열되고 용융 된 상태일 때 용융 된 수지는 약간의 점도를 가지는 유체이다. 그러나 용융수지는 금형 표면으로 열을 전달하기 때문에 스프루와 런너를 통과한 플라스틱이 캐비티에 도달하면 점도가 증가하기 시작한다.

and flowing becomes impossible thereafter.

Up to what distance the leading part can flow without cooling and solidifying? By knowing this it is possible to consider at the time of designing the mold the number and placement of gates, the placement of the runner, etc.

An index that becomes a guideline for that is the flow ratio (L/t). The flow ratio is an experimental index indicating the distance to which the leading edge of the flow can reach when a specific plastic is made to flow inside a cavity with a fixed plate thickness and at a fixed pressure.

The flow ratio is expressed, for example, as "the flow ratio (L/t) is 450 to 530 ㎜ when POM plastic is made to flow inside a cavity with a wall thickness of 1 ㎜ and with an injection pressure of 900 kgf/cm2".

In general, the following trends are shown by the flow ratio.

점도가 일정 한계 이상으로 증가하고, 냉각에 의해 유동 선단부가 고화되면 그 후 유동이 불가능해진다.

유동 선단부가 냉각되거나 고화하지 않고 유동할 수 있는 유동거리는 얼마일까? 금형설계 시에 유동거리를 알아야 게이트의 수와 위치, 런너의 위치 등의 설계가 가능하다.

그 지침이 되는 지표는 유동비 (L/t)이다. 유동비는 일정한 고정판 두께와 고정된 압력으로 특정 플라스틱을 내부로 흐르게 할 때 유동 선단이 도달 할 수 있는 거리를 나타내는 실험 지수이다.

예를 들면 이 유동비는「살 두께 1 ㎜, 사출압력 900 kgf/㎠으로 POM 용융수지를 캐비티에 유동 시켰을 때의 유동비 (L/t)는 450 ~ 530 ㎜ 이다.

일반적으로 다음과 같은 경향이 유동비로 표시된다.

1) The flow ratio value increases as the injection pressure increases.
2) The value decreases as the cavity plate thickness decreases.
3) The value increases as the cavity surface temperature increases.
4) The value shows some fluctuation depending on the condition of the molding machine and of the mold.
5) The flow state of a partially thin plate thickness part cannot be the target of prediction.

1) 유동비 값은 사출압력이 증가함에 따라 증가한다.
2) 캐비티 플레이트 두께가 감소함에 따라 유동비가 감소한다.
3) 캐비티 표면 온도가 증가함에 따라 유동비가 증가한다.
4) 성형기 및 금형의 상태에 따라 약간의 변동을 보인다.
5) 부분적으로 얇은 살 두께 부분의 유동 상태는 예측의 대상이 될 수 없다.

The flow ratios of major plastics are given below.

주요 플라스틱의 유동비는 다음과 같다.

Plastic name (kgf/cm²)	Injection pressure (mm)	Flow ratio (mm)	Cavity thickness (L/t)
POM Natural	900	1	450~530
ABS Natural	900	1	270~310
PBT with 30% glass fibe	1000	1	110~130

[Fig.]The flow ratio (L/t) of plastics

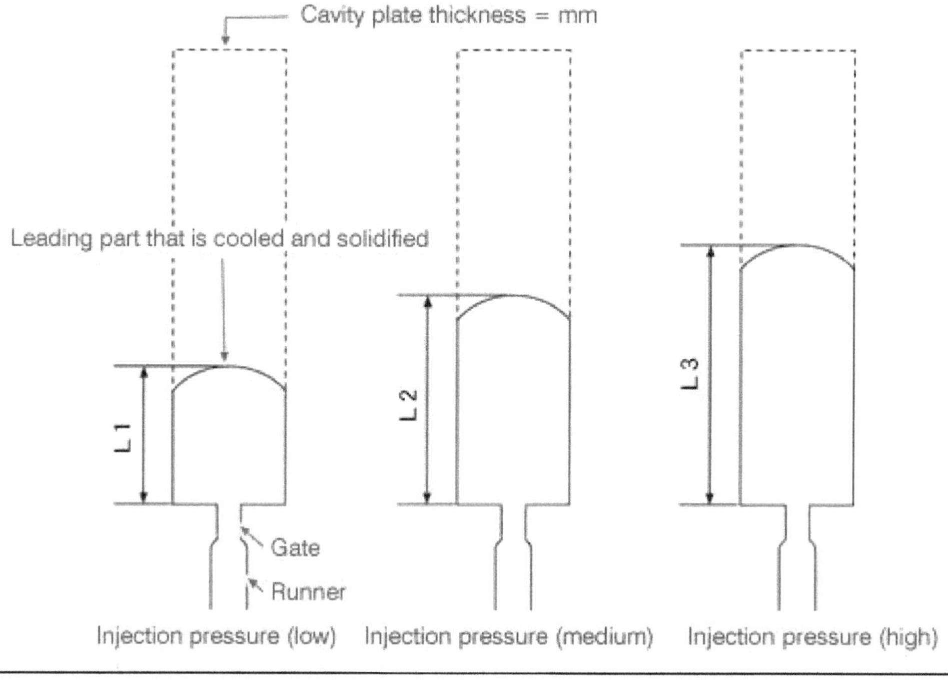

B 007 Method of Forming Holes in Molded Products

In this course, we explain the basic structure of a mold when a hole is to be prepared in a molded product. In order to prepare a hole in the molded product, it is necessary to form a part in to which plastic does not flow using the cavity or a core pin. An example of the basic structure for this is shown in [Fig. 1].

B-007 성형품의 구멍 성형 방법

이 과정에는 성형품에 구멍을 성형할 경우 금형의 기본 구조를 설명한다. 성형품에 구멍을 성형하기 위해서 캐비티 또는 코어 핀을 사용하여 용융 된 수지가 흐르지 못하게 한다. 이에 대한 기본 구조는 그림 1과 같다.

[Fig. 1] Example of a basic structure for preparing holes in a molded product

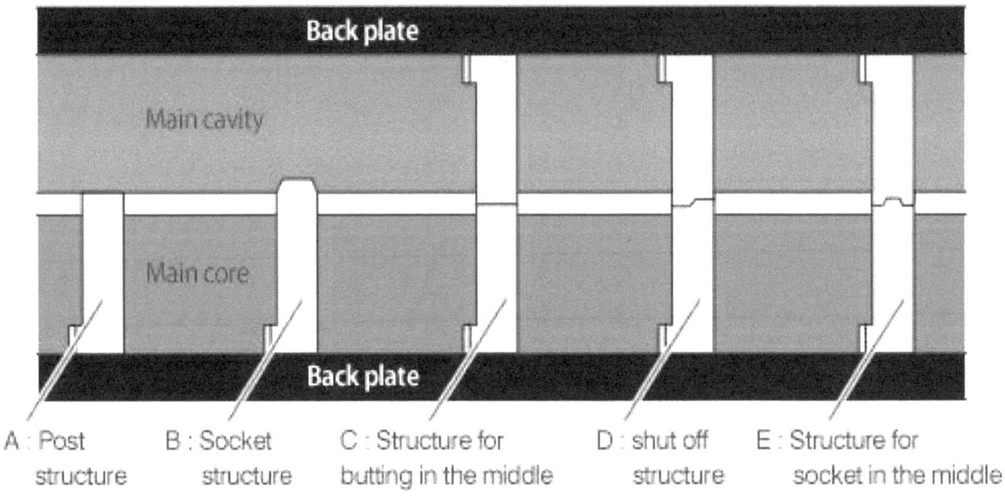

A. Touching structure

This is the most basic structure. A hole is formed by providing a projection from one side making it touch against the plane surface on the other side. A know-how, regarding the surface on the projection side that is touched, is, while preparing the core pins, to make them longer than the reference height by about 0.005 to 0.03 ㎜ as the "compression margin". If this is done, it will be difficult for flash to form on the touching surface, and sharp edges can be obtained in the periphery of the hole in the molded product. A drawback to this structure is that, when the core pin is thin and long, the core pin gets deformed due to the filling pressure of the plastic thereby causing the likelihood of shifts in the hole position, or a bend in the hole. In addition, in some cases the core pin may bend and break due to the filling pressure.

A. 접촉 구조

가장 기본적인 구조이다. 한 면에서 돌출 된 부분이 다른 면의 평면과 접촉되면서 성형된다. 이때의 노하우는 접촉되는 코어 핀의 제작시 압축률을 고려하여 약 0.005 내지 0.03 ㎜ 만큼 길게 만든다. 이렇게 되면 접촉면에 플래쉬의 발생과 구멍의 주변부가 날카로워 지는 것을 막을 수 있다. 이러한 구조의 결점은 코어 핀이 얇고 길 때 코어 핀이 사출압력으로 인하여 변형되어 홀 위치에서의 옮겨가거나 홀 내의 굴곡을 야기한다는 것이다. 또한, 경우에 따라 코어 핀이 사출압력으로 인하여 구부러지고 부서 질 수 있다.

B. Socket structure

In this structure, the tip of the projection provided from one side is made to engage in a hole provided on the other side thereby forming a hole. If this structure is used, since the core pin will have the structure of a beam that is supported at both ends, the capacity to resist the pressure of the plastic becomes better than that of the touching structure and has the effect of preventing bending and breaking of the core pin or shift in the hole position.

A taper is provided on the side periphery of the tip of the projection and the mating hole so that they can mate smoothly. A drawback to this structure is that the cost of preparing the mold will be higher than that of the touching structure.

B. 소켓 구조

이 구조는 금형의 한 측면으로부터 돌출 된 코어의 팁을 반대측 금형의 측면에 가공된 구멍에 결합하여 구멍을 성형한다. 이 구조를 사용하면 코어 핀이 양단에서 지지되는 빔 구조를 갖기 때문에 사출압력에 저항하는 능력은 접촉 구조의 저항 성능보다 우수하고, 굽힘 및 파손을 방지하는 효과를 갖는다. 돌출된 코어 끝의 측변부와 결합되는 구멍에 테이퍼가 만들어지므로 매끄럽게 결합된다. 이 구조의 단점은 금형을 가공하는 비용이 접촉 구조의 가공비용보다 비싸다는 점이다.

C. Structure for butting in the middle

In this structure, projections are provided from both sides so that they butt against each other in the middle. In this structure, since the total length of the core pins can be made shorter, it is possible to reduce the likelihood of breaking the core pin. A drawback to this structure is that a parting line is generated in the middle of the hole in the molded product.

C. 중간에서 맞닿는 구조

이 구조에서 코어 핀은 양측으로부터 삽입되어 중간에서 서로 맞닿는다. 이 구성은 코어 핀의 전체 길이를 짧게 할 수 있기 때문에 코어 핀을 파손 할 가능성을 감소시킬 수 있다. 이 구조의 단점은 성형품의 구멍의 중간에 파팅 라인이 생기는 것이다.

D. Shut off structure

In this structure, projections are provided from both sides, and their angles are matched in the middle.

D. 차단 구조

이 구조에서 코어 핀이 양쪽에서 삽입되며, 그 각도는 중간에서 일치한다.

E. Structure for socket in the middle

This structure is intermediate between the structures of B and C.

E. 중간의 소켓 구조

이 구조는 B와 C의 구조의 특징을 동시에 가지고 있다.

B 008 Molding Cycle and Cooling Time

B-008 성형 사이클과 냉각시간

While plastic injection molds are required to have the functions for producing molded parts with the desired quality, at the same time it is also required that production be possible at the lowest possible production cost.

플라스틱 사출 금형은 원하는 품질의 성형 부품을 생산하는 기능이 있어야 함과 동시에 가능한 최저 생산 비용으로 제작이 가능해야한다.

The cycle of plastic injection molding is defined as follows.

플라스틱 사출성형의 사이클은 아래와 같이 정의 된다.

Molding cycle t (sec) = t1 + t2 + t3 + t4,

where, t1 is the injection time = injection time + dwelling time (sec), t2 is the cooling time (sec), t3 is the time (sec) needed to remove the molded product, and t4 is the time (sec) needed to open and close the mold.

성형 사이클 t(sec) = t1 + t2 + t3 + t4,

여기서, t1은 사출시간 = 사출시간 + 보압시간(초), t2는 냉각시간(초), t3은 성형품을 취출하는 데 필요한 시간(초)이고, t4는 금형을 열고 닫는 데 필요한 시간(초) 이다.

Among the factors determining the molding cycle, the one that is the most important is the cooling time t2. Cooling time is the time from filling the inside of the cavity with molten plastic to the sealing of the gate until the plastic solidifies. From experience, it is known that the cooling time varies depending on the cooling capacity of the cavity of the mold. In addition, it also varies depending on the type of molding material and the wall thickness of the molded product. Predicting what the optimum cooling time is during the mold design stage is a very important matter in estimating the production cost of the molded product. While recently software products have come on the market that predict the cooling time by CAE, in general, the following experimental equation is used for predicting the cooling time.

성형 사이클을 결정하는 요인 중 가장 중요한 것은 냉각시간 t2이다. 냉각시간은 캐비티 내부를 용융수지로 채우고 용융수지가 고화될 때까지 게이트를 밀폐하기 위하여 필요한 시간이다.

경험에 의하면 냉각시간은 캐비티의 냉각 용량에 따라 달라지는 것으로 알려져 있다. 또한, 성형 재료의 종류 및 성형품의 살 두께에 따라서도 다르다. 금형설계 단계에서 최적의 냉각시간이 어느정도 예측하는 것은 성형 된 제품의 생산 원가를 추정 할 때 매우 중요한 문제이다. 최근에 CAE에 의한 냉각시간을 예측하는 소프트웨어 제품이 출시되었지만 일반적으로 냉각시긴을 예측하기 위해 다음 실험식이 사용된다.

$t_{la} = s^2/(\pi^2 \cdot \alpha)\ \ln(8/\pi^2 \cdot (\theta r - \theta m)/(\theta e - \theta m))$, where, t_{la} is the cooling time (sec) related to the average temperature of the wall thickness; s is the wall thickness (㎜) of the molded product; α is the heat diffusion rate of the plastic at the cavity surface temperature (㎟/sec), $\alpha = \lambda/(c\rho)$; λ is the coefficient of thermal conductivity of the plastic (kcal/m • h • ℃); c is the specific heat of the plastic (kcal/kg • ℃); ρ is the density of the plastic (kg/m³); θr is the temperature of the molten plastic (℃); θe is the temperature for taking out the molded product (℃); and θm is the cavity surface temperature (℃).

* Reference: "Molds for Injection Molding" by Keizo Mitani, Sigma Publications, 1997 (in Japanese)

$t_{la} = s^2/(\pi^2 \cdot \alpha)\ln(8/\pi^2 \cdot (\theta r - \theta m)/(\theta e - \theta m))$, t_{la}는 살 두께의 평균 온도와 관련된 냉각시간(초)이다. s는 성형품의 두께 (㎜), α는 캐비티 표면 온도에서 플라스틱의 열확산계수(㎟/sec), α=λ/(C • ρ), λ는 플라스틱의 열전도계수(kcal/m • h • ℃), C는 플라스틱의 비열(kcal/kg • ℃), ρ는 플라스틱의 밀도(kg/m³), θr는 플라스틱의 용융온도(℃), θe는 성형품의 취출온도(℃), θm는 캐비티 표면 온도(℃).

*참고문헌 : Molds for Injection Molding" by Keizo Mitani, Sigma Publications, 1997 (in Japanese)

B 009 Flow Rate of Mold Cooling Water

In order to control the temperature of the mold, it is common to use water as the coolant if the temperature is less than 100℃. The cooling water whose temperature is controlled by the re-circulating pump of the temperature controller, circulates inside the cooling water holes provided inside the mold, and stabilizes the temperature of the mold by heat transfer and radiation.

Normally, it is not possible to see from the outside of the mold the way that the cooling water is circulating. In actuality there are two patterns for the flow of cooling water in terms of fluid mechanics, namely, "laminar flow" and "turbulent flow".

B-009 금형 냉각수의 흐름율

금형의 온도를 제어하기 위해 온도가 100°C 미만인 경우 냉각수로 물을 사용하는 것이 일반적이다. 온도 제어기의 재순환 펌프에 의해 온도가 제어 된 냉각수는 금형 내부에 설치된 냉각수 구멍을 순환하여 열전달 및 방사에 의해 금형의 온도를 안정시킨다.

일반적으로 냉각수가 순환하는 방식을 금형 외부에서 볼 수 없다. 실제로 유체 역학의 측면에서 냉각수의 흐름에 대한 두 가지 패턴, 즉 "층류"와 "난류"가 있다.

In order to efficiently carry out temperature control of the mold, the desirable condition for the flow of the cooling water is a "turbulent flow". The condition to make the flow turbulent can be roughly calculated by the indices called the dynamic coefficient of the viscosity of the fluid, the hole diameter, and the Reynolds number that is determined by the flow speed.

Limiting the discussion only to water, it is possible to realize a turbulent flow of the cooling water by making the water flow at more than a certain speed. In other words, if the diameter of the cooling water hole of the mold is determined, in order to make the flow turbulent, it is sufficient to make the supply flow rate from the circulating pump higher than a certain value. In this case, since the dynamic coefficient of viscosity changes with the water temperature, it is necessary to change the flow rate in proportion to the water temperature.

The data that are indices for a turbulent flow are listed in the following Table.

Table:

Turbulent flow region limiting flow rate of cooling water (L/min)

금형의 온도 제어를 효율적으로 수행하기 위하여 바람직한 냉각수의 흐름은 "난류"이다. 유동을 난류로 만드는 조건은 유체의 동점도 계수, 냉각수 홀의 직경, 그리고 유속에 의해서 정의되는 레이놀드 수 (Reynolds number)에 의해서 대략적으로 계산할 수 있다.

냉각수에 한정해서 특정 속도 이상에서 물 흐름을 만들어 냉각수의 난류 유동을 실현 할 수 있다. 다른 말로 유동흐름을 난류로 만들기 위한 금형의 냉각수 구멍 직경이 정의 된다면 특정 값보다 더 큰 순환 펌프로부터 유량을 공급하므로서 충분히 가능하다.

이 경우 동점성 계수는 물의 온도와 같이 변화하기 때문에 물 온도와 비례하여 유량을 변경할 필요가 있다.

난류 흐름에 대한 지표 데이터는 다음 표에 나열되어있다.

테이블 :

냉각수의 난류 영역 제한 유량 (L/min)

	Cooling water hole diameter (㎜)			
Water temperature (℃)	6	8	10	14
20	0.99	1.32	1.62	2.3
40	0.65	0.84	1.08	1.5
60	0.48	0.63	0.80	1.1
80	0.36	0.48	0.47	0.83
100	0.37	0.48	0.47	0.83

B 010 Example of Calculating the Cooling Time (1)

B-010 냉각시간 계산 예 (1)

Problem:

How long is the necessary cooling time for an injection molded product made of ABS plastic (natural material) with a wall thickness of 1.5 ㎜? However, assume that the cavity surface temperature is 50℃, the temperature of the molten plastic is 230℃, and the molded product releasing temperature is 90℃.

예제 :

살 두께가 1.5 ㎜인 ABS 플라스틱(천연 소재)으로 만들어진 사출성형 제품에 필요한 냉각시간은 얼마인가? 단, 캐비티 표면 온도가 50℃, 수지 용융온도가 230℃, 성형품 취출 온도가 90℃라고 가정한다.

Sample answer:

The cooling time t_{la} required until the average temperature of the molded product becomes 50℃ is calculated using the following equation.

$$t_{la} = s^2/(\pi^2 \cdot \alpha)\ln(8/\pi^2 \cdot (\theta r - \theta m)/(\theta e - \theta m))$$

where, t_{la} is the cooling time(sec) related to the average temperature of the wall thickness; s is the wall thickness 1.5 (㎜) of the molded product; α is the heat diffusion rate of the plastic at the cavity surface temperature, $\alpha = \lambda/(c \cdot \rho)$; λ is the coefficient of thermal conductivity of the plastic (kcal/m • h • ℃); c is the specific heat of the plastic (kcal/kg • ℃); ρ is the density of the plastic (kg/m^3); and with the ABS plastic (natural material) at a cavity surface temperature of 50℃, α = 0.0827 ㎜^{2}/sec; θr is the temperature of the molten plastic (230℃); θe is the temperature for taking out the molded product (90℃); and θm is the cavity surface

예시 답안 :

성형품의 평균 온도가 50℃가 될 때까지 필요한 냉각시간 (t_{la})은 다음 식을 사용하여 계산된다.

$$t_{la} = s^2/(\pi^2 \cdot \alpha)\ln(8/\pi^2 \cdot (\theta r - \theta m)/(\theta e - \theta m))$$

여기서, t_{la}는 살 두께의 평균 온도와 관련된 냉각시간(초)이다. s는 성형품의 살두께 1.5 (㎜)이고, α는 캐비티 표면 온도에서 플라스틱의 열 확산율이며, $\alpha = \lambda/(c \cdot \rho)$; λ는 플라스틱의 열전도계수 (kcal/m • h • ℃)이다. c는 플라스틱의 비열 (kcal/kg • ℃); ρ는 플라스틱의 밀도 (kg/m^3); 캐비티 표면 온도가 50℃인 ABS 플라스틱 (무첨가제)의 경우, α = 0.0827 ㎜^{2}/sec; θr은 용융수지의 온도 (230℃); θe는 성형품 취출 온도 (90℃); θm은 캐비티 표면 온도 (50℃)이다.

temperature (50℃). Substituting these values into the above equation, we get:

$$t_{la} = 1.5^2/(3.14^2 \cdot 0.0827)\ln(8/3.14^2 \cdot (230-50)/(90-50)) = 3.57\,(\mathrm{sec})$$

The cooling time required until the average temperature of the molded product becomes 50℃ is 3.57 sec.

* Reference:"Molds for Injection Molding" by Keizo Mitani, Sigma Publications, 1997 (in Japanese)

위의 방정식에 이 값을 대입하면 다음과 같이 계산할 수 있다.

$$t_{la} = 1.5^2/(3.14^2 \cdot 0.0827)\ln(8/3.14^2 \cdot (230-50)/(90-50)) = 3.57(\text{초})$$

성형품의 평균 온도가 50℃가 될 때까지의 냉각시간은 3.57 초이다.

*참고문헌

B 011 Cooling Time of the Central Part of the Molded Article

While in the last lesson we discussed the experimental equation that predicts the cooling time for the average temperature of the wall thickness of the molded item to be cooled up to a prescribed temperature, in this lesson we explain the experimental equation that predicts the time taken for the temperature of the central part of the molded product to be cooled to a prescribed temperature.

When we want to reduce the so called "post-shrinkage" which is the shrinkage after the molded item has been taken out of the mold, it is necessary to cool the molded article inside the mold until the central part of the wall thickness of the molded item is cooled to a prescribed temperature.

B-011 성형품의 중심부까지 냉각시간

시난 강의에서 우리는 성형품의 평균온도가 정해진 온도까지 도달할 때까지의 시간을 예측하는 실험식을 논의한 바 있다. 이 강의에는 성형품의 중심부 온도가 정해진 온도로 냉각되는 데 걸리는 시간을 예측하는 실험식을 설명한다.

성형품을 금형에서 꺼낸 후 수축되는 소위 "후 수축"을 줄이려면 성형품의 중앙부가 정해진 온도로 냉각 될 때까지 금형 내부에서 성형품을 냉각 할 필요가 있다.

The cooling time t_{lc} for the central part of the wall thickness of the molded article to reach the temperature θe can be calculated using the following equation.

$$t_{lc} = s^2/(\pi^2\alpha)\ln(4/\pi \cdot (\theta r - \theta m)/(\theta e - \theta m))$$

t_{lc} : The cooling time (seconds) for the central part of the wall thickness of the molded article to reach the temperature θe.

Where,

s : Wall thickness of the molded article (㎜)

α : Heat diffusion ratio of the plastic at the cavity surface temperature (㎜²/sec)

$\alpha = \lambda/(c \cdot \rho)$

λ = Thermal conductivity of the plastic resin (kcal/m • h • ℃)

c : Specific heat of the plastic resin(kcal/kg • ℃)

ρ : Density of the plastic resin(kg/m³)

θr : Temperature of the molten plastic resin (℃)

θe : Temperature at the center of the wall thickness of the molded article (℃)

θm: Cavity surface temperature (℃)

* References: "Molds for injection molding" (Keizo MITANI, Sigma Publications (1997))

성형품의 중앙부 온도가 θe에 도달할 때까지의 냉각시간(t_{lc})은 다음 식을 사용하여 계산 할 수 있다.

$$t_{lc} = s^2/(\pi^2\alpha)\ln(4/\pi \cdot (\theta r - \theta m)/(\theta e - \theta m))$$

t_{lc} : 성형품의 중앙부 온도가 θe에 도달하기까지의 냉각시간(초).

여기서,

s : 성형품의 살 두께 (㎜)

α : 캐비티 표면 온도에서의 플라스틱의 열 확산율 (㎜²/sec)

$\alpha = \lambda/(c \cdot \rho)$

λ = 플라스틱 수지의 열전도율 (kcal/m • h • ℃)

c : 플라스틱 수지의 비열 (kcal/kg • ℃)

ρ : 플라스틱 수지의 밀도 (kg/m³)

θr : 용융수지의 온도 (℃)

θe : 성형품 중심부의 온도 (℃)

θm: 캐비티 표면 온도 (℃)

* 참고문헌 : "Mold for injection molding" (Keizo MITANI, Sigma Publications (1997))

B 012 Pressure Loss in the Runners

B-012 런너에서의 압력손실

In plastic injection molding, the molten plastic material flows through the runners, passes through the gates, and then reaches the cavity. The pressure gradually decreases in the process of this sequence of flow. Molten plastic is a viscous fluid having a certain amount of viscosity. In addition, it also has the feature that the viscosity changes depending on the temperature of the plastic, and when the plastic temperature falls below a certain range the plastic can no longer flow and starts to solidify.

Further, whenever a viscous fluid flows through a flow path, there is always a loss in pressure. This is the same as with the flow of water or oil. Now, under what situations is the pressure loss high? Pressure loss is known to occur under the following conditions.

플라스틱 사출성형에서 용융 된 플라스틱 재료는 런너를 통과하여 게이트를 지나서 캐비티에 도달한다. 이 과정에서 압력은 점차 감소한다. 용융 된 플라스틱은 일정한 점도를 갖는 점성유체이다. 또한 점도는 플라스틱의 온도에 따라 달라지며 플라스틱 온도가 일정 범위까지 하락하면 용융수지는 더 이상 흐르지 않고 응고되기 시작한다.

또한, 점성유체가 유동경로를 통해 유동 할 때마다 항상 압력손실이 있다. 이것은 물이나 기름의 유동과 같다. 그러하면 어떤 상황에서 압력손실이 높아질까?

압력손실은 다음 조건에서 발생하는 것으로 알려져 있다.

(1) In the vicinity of the inlet to the flow path

Whirls are generated in the vicinity of the inlet to the flow path into which the fluid enters thereby causing a pressure loss. It is possible to reduce the pressure loss by making the rounding diameter R large for the corners of the inlet.

(1) 유로의 입구 부근

유체가 유입되는 유로의 입구 근방에 와류가 발생하여 압력손실이 발생한다. 입구 모서리의 둥근 직경 R을 크게 함으로써 압력손실을 감소시킬 수 있다.

(2) Where the flow path bends

A pressure loss occurs in the parts where the flow path bends at an angle because there is a change in which the fluid first gets compressed and then expands.

(2) 구부러진 유동경로

유체는 먼저 압축된 후 팽창하는 변화가 있기 때문에 유체가 비스듬히 굴절하는 부분에서 압력손실이 발생한다.

(3) Where flexures and bends are present

Whirls are generated in parts where there are flexures and bends thereby causing a loss in pressure.

(3) 휘어짐과 굽힘이 있는 곳

압력손실의 원인이 되는 휘어짐과 굴곡이 있는 제품부에서 와류가 발생한다.

(4) Locations where the flow path expands or contracts

Whirls are generated in parts where the cross section of the flow path becomes wider or narrower thereby causing a loss in pressure. In particular, since a large pressure loss occurs with very sudden expansions or contractions of the flow path, it is necessary that they be strictly avoided.

(4) 유로가 확장되거나 축소되는 위치

유동경로의 단면이 넓어지거나 좁아지는 부분에서 와류가 발생하여 압력손실이 발생한다. 특히, 유로의 급격한 팽창 또는 수축으로 큰 압력손실이 발생하기 때문에 엄격히 피해야 한다.

Although it is possible to know the trends in pressure loss due to the predictions in pressure loss using CAE tools, in order to optimize the detailed flow conditions that become necessary in actual injection molding, it is very important to understand the reasoning according to theory and to utilize the know how obtained by the trial and error of fine adjustments.

실제 사출성형에 필요한 자세한 유동 조건을 최적화하기 위해 CAE를 이용하여 압력손실을 예측하여 압력손실의 추세를 알 수 있지만 이론에 따라 추론 후 이해하는 것이 매우 중요하다. 시행착오를 통해 얻은 노하우를 미세 조정으로 활용할 수 있다.

B 013 Thermal Expansion of Molds

B-013 금형의 열팽창

In the case of plastic injection mold, since it is necessary to maintain the temperature of the mold within a range of 20 to 200℃, and since there will be a difference between this temperature and room temperature, there will be thermal expansion of the parts of the mold (in some cases, there may be thermal contraction).

플라스틱 사출성형의 경우, 금형의 온도를 20 ~ 200℃의 범위 내로 유지할 필요가 있고, 유지 온도와 실온 사이에 차이가 있기 때문에 금형의 일부(경우에 따라 열 수축이 있을 수 있음)에 열팽창이 발생한다.

Although it is very common to carry out the adjustment of the assembly of a mold at room temperature in a constant temperature room, the temperature region differs during actual injection molding.

Each part of a mold has its own shape, and since a number of such parts are assembled in a complex manner, there will be sliding or friction movement.

Even the state of thermal expansion is extremely complex and it is very difficult to predict it by theory.

The practical method of using know-how accumulated by experience is very common regarding the problem of thermal expansion.

The method of supplying oil is modified, the contact surface area is reduced, and the amount of deformation is absorbed by providing an escape.

In addition, a technique is used of balancing the thermal bias by using additionally the temperature controlling function. A cooling water flow path is provided near the heater or the manifold, or an air thermal insulation layer is provided, or a thermal insulation plate is provided.

Positioning rails are placed in the case of sliding parts.

Ball bearings are used in the case of parts lubricating which is difficult, or even a non-lubricating type alloy is used.

In order to balance the heat accumulation in the mold base, the excel thickness of unwanted parts are removed.

Regarding hot runners, it is particularly important to calculate and study thoroughly the thermal balance between heat supply and loss. It is necessary to give proper considerations about the balance between the quantity of heat on the fixed side and the quantity of heat on the movable side of the chip parts, manifold parts, etc.

상온이 일정하게 유지되는 공간에서 금형 조립을 조정하는 것은 매우 일반적인 일이지만 실제 사출 성형시의 온도와는 차이가 있다.

금형의 각 부분은 각각의 모양을 가지며, 여러 부분이 복잡한 방식으로 조립되기 때문에 슬라이딩 또는 마찰 운동이 발생한다.

열팽창의 상태조차도 매우 복잡하며 이론적으로 예측하기가 매우 어렵다.

경험에 의해 축적된 노하우를 사용하는 실용적인 방법은 열팽창 문제와 관련하여 매우 일반적이다.

오일을 공급하는 방법을 수정하고, 접촉면을 줄이고, 탈출구를 제공하여 변형량을 흡수한다.

또한, 온도 제어 기능을 추가로 사용하여 열 편중을 균형적으로 하는 기술이 사용된다. 히터 또는 매니 폴드의 근방에 냉각수 유로가 설치되거나, 공기 단열층이 설치되거나 단열판이 설치된다.

포지셔닝 레일은 슬라이딩 부에 배치된다.

볼 베어링은 윤활이 어려운 부품의 경우 또는 비 윤활 타입의 합금을 사용하는 경우에 적용된다.

몰드베이스의 열 축적을 균형적으로 하기 위해 원하지 않는 과도한 두께를 제거한다.

핫 런너에 관해서는 열 공급과 손실 사이의 열 균형을 철저히 계산하고 연구하는 것이 특히 중요하다. 고정측 금형의 열량과 칩 부품, 매니폴더 부품, 기타 부품 등이 있는 가동측 금형의 열량사이의 균형에 관해서는 적절한 고려가 필요하다.

It is also necessary to consider the amount of heat radiated from the platen of the molding machine.

또한 성형기의 설치판에서 방사되는 열량을 고려해야한다.

B 014 Molding Conditions of an Injection Molding Machine (1)

B-014 사출성형기의 성형 조건 (1)

1. Molding Conditions of an Injection Molding Machine

In order to form a molded article by actually carrying out injection molding using a plastic injection molding machine, it is necessary to carry it out in a condition in which the mold and the molded material are made to match with the specifications of the molded article. If merely a molten plastic is injected into the mold, solidified, and taken out of the mold, there will be large variations in the molded article dimensions, fluctuations in glossiness, and generation of external appearance transfer striations, and it is not possible to obtain products with a uniform quality.

In order to obtain injection molded articles with the desired quality, it is necessary to adjust various parameters called the "molding conditions" of the injection molding machine by carrying out the work of adjusting the conditions so that the specifications are satisfied while controlling the dimensions, external appearance, and quality of the molded article.

In other words, the quality of the produced injection molded article depend on whether the adjustment of the molding conditions is good or bad. If the range of this parameter adjustment of this adjustment of molding conditions is wide, the range of changing the quality specifications becomes wide, and making the adjustments

1. 사출성형기의 성형 조건

사출성형기를 이용하여 실제 성형품을 성형하기 위해서는 성형품의 사양에 맞는 금형과 재료 조건하에서 성형해야 한다. 단순히 용융수지를 금형에 주입하여 고화시켜 금형 밖으로 취출 한다면 성형품 치수에 큰 편차, 광택도의 변동 및 외관에 줄무늬가 있을 수 있어서 균일한 품질의 제품을 얻는 것이 불가능하다.

원하는 품질의 성형품을 얻기 위해서 사출성형기의 "성형 조건"이라 불리는 각종 파라미터를 조정하는 동시에, 성형품의 치수, 외관를 제어하면서 사양을 만족하도록 조건을 조정하는 작업이 필요하다.

즉, 성형된 사출성형품의 품질은 성형 조건의 양호한 조정의 여부에 따라 결정된다. 성형 조건의 조정 범위가 넓으면 품질 사양의 변경 범위가 넓어지므로 조정이 용이해 진다.

becomes easy. When the usual comment that "the mold has been prepared well" is made, very often the width of such condition adjustment is wide. To put it in another way, during the design of the mold, it is necessary to give considerations so that it is possible to set wide the range of adjustment of the molding conditions. For example, the design is made so that sufficient flow rate of the cooling circuit is acquired, providing gas vents appropriately thereby making it difficult for burns to occur even when high speed filling is made, etc.

The typical molding condition parameters are explained below.

평소에 "금형이 잘 준비되었다"라는 말은 성형 조건의 조정 범위가 매우 넓음을 의미한다. 달리 말하자면, 금형설계 중에 성형 조건의 조정 범위를 넓게 설정할 수 있도록 고려할 필요가 있다. 예를 들어, 냉각 회로의 충분한 유속이 얻어 지도록 설계되고, 가스 배출구를 적절히 제공함으로써 고속 충전을 하더라도 성형품이 타는 문제가 발생하지 않게 한다.

일반적인 성형 조건 변수는 아래에 설명되어 있다.

(1) Barrel temperature

The barrel temperature can be adjusted by the temperature setting of the injection barrel. The appropriate range of the barrel temperature can be set while referring to the recommended temperature range that varies with the grade of the resin and which is provided by each resin manufacturer.

It is possible to measure the actual resin temperature by inserting the thermocouple of a thermocouple type thermometer into the resin that has been injected in a purging injection.

(1) 배럴 온도

배럴 온도는 사출성형기 배럴의 온도 설정에 따라 조정할 수 있다. 배럴 온도의 적절한 범위는 수지의 그레이드 따라 다르며 각 수지 제조업체가 제공하는 권장 온도 범위를 참조하여 설정한다.

퍼지 주입으로 주입된 수지에 서머커플 타입의 온도계를 수지에 삽입하여 실제 수지 온도를 측정 할 수 있다.

(2) Mold temperature

The mold temperature is a very important parameter that affects the shrinkage of the molded article and the transfer of the cavity surface. It is very common to use a water cooled type temperature controller when the mold temperature is in the range of 15 to 90℃. When the mold temperature exceeds 90℃, either a pressurized water type or an oil type temperature controller is used,

(2) 금형 온도

금형 온도는 성형품의 수축 및 캐비티 표면의 전사에 영향을 미치는 매우 중요한 변수이다. 금형 온도가 15 ~ 90℃ 일 때 수냉식 온도 컨트롤러를 사용하는 것이 일반적이다. 금형 온도가 90℃를 초과하면 가압수식 또는 오일식 온도 제

or the temperature control is made using a cartridge heater.

Although what is being controlled is the mold temperature, what is really important is the cavity surface temperature. In order to control this accurately, the common method is to adjust the mold temperature based on the actual value measured using a contact type surface temperature measuring instrument.

As for the method of incorporating a non-contacting type infrared temperature sensor inside the cavity, although it is very effective because it is possible to measure accurately the mold temperature in real time, it requires equipment and know how.

어기가 사용되거나, 카트리지 히터를 사용하여 온도 제어가 이루어진다.

제어되는 것은 금형 온도이지만, 실제로 중요한 것은 캐비티 표면 온도이다. 이를 정확하게 제어하기 위해서 접촉식 표면 온도 측정기를 사용하여 측정한 실제 값을 기준으로 금형 온도를 조정하는 것이 일반적이다.

캐비티 내에 비접촉식 적외선 온도 센서를 내장하는 방법은 실시간으로 금형 온도를 정확하게 측정할 수 있기 때문에 매우 효과적이지만 장비와 노하우가 필요하다.

B 015 Molding Conditions of an Injection Molding Machine (2)

B-015 사출성형기의 성형 조건 (2)

1. Filling Pressure

Among the plastic injection molding conditions, the filling pressure, also called the primary pressure, is the pressure required for filling the inside of the mold with plastic. The unit of filling pressure is MPa or kgf/cm^2. That is, it is the unit of pressure. There are also injection molding machines that adjust the condition by changing what percentage of the maximum filling pressure of the machine is to be used as the filling pressure (for example, 35%).

1. 충전압력

플라스틱 사출성형 조건 중, 1차 압력이라고도 하는 충전압력은 금형 내부를 플라스틱으로 채우는 데 필요한 압력이다. 충전압력 단위는 MPa 또는 kgf/cm^{2}로 표시된다. 즉, 압력의 단위이다. 사출성형기의 최대 충전압력의 몇 퍼센트를 충전압력(예 : 35%)으로 사용하여 조건을 설정하는 사출성형기도 있다.

Filling defects (short shot) occurs when the filling pressure is too low. When the filling pressure is too high, the parting surface opens instantaneously due to the pressure, which may cause flashes (burrs) to appear on the periphery of the molded item. Further, there may be defective releasing of the molded item from the mold.

충전 불량(미성형)은 충전 압력이 너무 낮을 때 발생한다. 충전압력이 너무 높으면 압력으로 인해 순간적으로 파팅면 열리면서 성형품 주변부에 버(burr)가 생길 수 있다. 또한, 성형품의 취출 불량이 발생할 수 있다.

2. Injection Speed

The injection speed is the speed of the molten plastic when it is being injected into the mold. The unit of this speed is ㎜/s.

This is also indicated as the injection rate which is the injected volume per unit time. The unit in this case will be cm^3 /s.

Filling defects can occur when the injection speed is too low, and burrs and jetting (the defect in which corrugated pattern appears on the surface of the molded item) can occur when the filling speed is too high.

2. 사출속도

사출속도는 용융수지가 금형에 주입 될 때의 속도이다. 이 속도의 단위는 ㎜/s이다.

이것은 또한 단위 시간당 주입된 부피를 사출률로 표시된다. 이 경우 단위는 cm^3/s이다.

사출속도가 너무 느리면 충전불량이 발생할 수 있으며, 사출속도가 너무 빠르면 버 및 젯팅(성형품 표면에 주름이 나타나는 결함)이 발생할 수 있다.

3. Dwelling Pressure

The dwelling pressure is the pressure that is applied during the period after the entire cavity has been filled with molten plastic until the gate has been sealed (the gate has solidified) so that the plastic does not flow back from the gate. Its unit is MPa or kgf/cm^2 .

When the dwelling pressure is too low, the plastic flows back from the gate causing the surface of the molded article to get dented (sink), or the shrinkage becomes too large making the dimensions smaller.

3. 보압

보압은 전체 캐비티가 용융수지로 채워진 후 용융수지가 게이트에서 다시 흘러나오지 않도록 게이트가 밀봉될 때까지 (게이트가 응고될 때까지) 적용되는 압력이다. 단위는 MPa 또는 kgf/cm^2 이다.

보압이 너무 낮으면 용융수지가 게이트에서 흘러 나와 성형품의 표면이 움푹 들어가거나 수축이 너무 커서 치수가 작아진다.

When the dwelling pressure is too high, the dimensions may become too large, or burrs can occur, or there may be defective releasing from the mold.

The dwelling time is very important in making the dwelling pressure have the appropriate effect, and it is very important to understand that the dwelling pressure must not be released until the gate has been sealed.

보압이 너무 높으면 치수가 너무 커지거나 버가 발생할 수 있으며, 금형에서 취출이 재대로 되지 않을 수도 있다.

보압시간은 보압이 적절한 효과를 발휘할 수 있게 하는데 중요하며, 보압은 게이트가 밀봉될 때까지 해제되어서는 안됨을 이해하는 것이 매우 중요하다.

4. Injection Time

Generally, the sum of the fill time and the dwelling time is called the injection time. The dwelling time is adjusted by changing the injection time. The unit is s (seconds).

4. 사출시간

일반적으로 충전시간과 보압시간의 합을 사출시간이라고 한다. 보압시간은 사출시간을 변경하여 조정된다. 단위는 s(초)이다.

B 016 Molding Conditions of an Injection Molding Machine (3)

B-016 사출성형기의 성형 조건 (3)

1. Cooling Time

The cooling time in plastic injection molding is the period of time over which the molded article is left in the mold to solidify. The cooling time starts when the dwell time ends. The unit is s (seconds).

If the cooling time is too short, the shrinkage becomes high making the dimensions small, and the molded article may also get deformed during its ejection from the mold.

1. 냉각시간

플라스틱 사출성형에서 냉각시간은 성형품이 금형에 남아 고화되는 시간이다. 보압시간이 끝나면 냉각시간이 시작된다. 단위는 초이다.

냉각시간이 너무 짧으면 수축이 커져서 치수가 작아지고, 금형으로부터 취출되는 동안 성형품이 변형될 수도 있다.

On the other had, if the cooling time is too long, the molding cycle becomes longer reducing the productivity. In other words, the cost of producing a molded article increases. Since the cooling time is a factor having the highest effect among all the factors affecting the molding cycle time, shortening the cooling time becomes a very important key to shortening the molding cycle time.

2. Measuring Position

The measuring position is the position of the screw or plunger for measuring the volume of plastic necessary for injection. The unit is ㎜.

Fill defects will occur if the measuring position is too short. If the measuring position is too long, excess plastic will remain in the cylinder causing burning, or generating gas. The measuring position should be set to the appropriate value considering the shot volume required for the molded item so that it is neither too small nor too large.

3. Speed - Dwell Switchover Position

This is the screw position for switching between the injection speed and dwell, and its unit is ㎜. When this position is changed, there will be fine changes in the external appearance and the dimensions of the molded article.

반면 냉각시간이 너무 길면 성형 사이클이 길어져 생산성이 저하된다. 즉, 성형품의 생산 비용이 상승한다. 냉각시간은 성형 사이클 타임에 영향을 미치는 모든 요소 중에서 가장 큰 영향을 미치는 요소이기 때문에 냉각시간을 단축하는 것이 성형 사이클 타임을 단축시키는 매우 중요한 열쇠이다.

2. 계량 위치

계량 위치는 사출에 필요한 플라스틱 체적을 측정하기 위한 스크류 또는 플런저의 위치이다. 단위는 ㎜이다.

계량 위치가 너무 짧으면 충전불량이 발생한다. 계량 위치가 너무 길면 과도한 플라스틱이 실린더에 남아 연소를 일으키거나 가스를 발생시킨다. 계량 위치는 성형품에 필요한 충전량을 고려하여 적절한 값으로 설정해야한다. 너무 작거나 너무 크지 않도록 해야 한다.

3. 속도 제어-압력 제어의 변화 위치(보압절환 위치)

이것은 사출과정에서 보압과정으로 절환을 위한 스크류 위치이며, 단위는 ㎜ 이다. 이 위치가 변경되면 성형품의 외관 및 치수가 미세하게 변경된다.

4. Screw Forward Position

This is the parameter that controls the position to which the screw advances to the most forward position after injection is completed. Its unit is ㎜.

This is also a factor that controls the amount of cushion. The amount of cushion is the molten plastic remaining inside the cylinder at the tip of the screw after the screw has moved to the most forward position. If the amount of cushion is small, the dwell pressure will not act properly.

If the amount of cushion is too much, the material remains and gets the effect of heat thereby deteriorating.

4. 스크류 선단 위치

이것은 사출이 완료된 후 스크류가 가장 앞쪽으로 전진할 때, 그 위치를 제어하는 변수이다. 단위는 ㎜이다.

이것은 또한 쿠션의 양을 제어하는 요소이다. 쿠션의 양은 스크류가 가장 앞쪽으로 이동 한 후 스크류 팁에서 실린더 내부에 남아있는 용융수지이다. 쿠션 량이 적으면 압력이 적절하게 작용하지 않는다.

쿠션의 양이 너무 많으면 재료가 남아 열의 영향을 받아 열화(劣化)된다.

B 017 Molding Conditions of an Injection Molding Machine (4)

B-017 사출성형기의 성형 조건 (4)

1. Screw Rotation Speed

The screw rotation speed in plastic injection molding is the speed of rotations of the screw for mixing the pellets. Its unit is rpm (revolutions per minute).

If the screw rotation speed is too high, air gets mixed inside the molten plastic which can make gas generation to occur easily. Also, if the screw rotational speed is too low, sufficient kneading will not be made and the material quality can fluctuate.

1. 스크류 회전속도

플라스틱 사출성형에서의 스크류 회전속도는 펠릿을 혼합하기 위한 스크류의 회전속도이다. 단위는 rpm (분당 회전 수)이다.

스크류 회전속도가 너무 빠르면 용융수지 내부에서 공기가 혼합되어 가스가 쉽게 발생할 수 있다. 또한, 스크류 회전속도가 너무 느리면 충분한 혼련이 이루어지지 않고 재료 품질이 변할 수 있다.

2. Screw Back Pressure

The screw back pressure is the pressure at the time that the screw moves back for measuring the material. Its unit is MPa or kgf/cm^2 . The kneading state of the material changes due to changes in the back pressure.

2. 스크류 배압

스크류 배압은 재료를 계량하기 위해 스크류가 뒤로 움직일 때의 압력이다. 단위는 MPa 또는 kgf/cm^2 이다. 재료의 용융 상태는 배압의 변화로 인해 변화한다.

3. Mold Opening and Closing Speed

The mold opening and closing speed is the speed at which the mold is opened and the speed at which the mold is closed. The unit is ㎜/s. Although the time required for opening and closing the mold being small is effective in shortening the cycle time, if the opening and closing speed is too high, the danger becomes high of the molds colliding with each other, and hence it is necessary to put a brake just before the molds meet.

Also, if the mold opening speed is too high, the mold releasing state fluctuates which can cause fluctuations in the quality.

3. 금형 개폐속도

금형 개폐속도는 금형이 열리는 속도와 금형이 닫히는 속도이다. 단위는 ㎜/s 이다. 금형의 개폐에 필요한 시간이 짧아도 사이클 타임의 단축에는 효과적이지만, 개폐속도가 너무 빠르면 금형끼리 충돌하는 위험성이 높아지기 때문에 금형이 만나는 직전에 브레이크를 작동시켜야 한다.

또한, 형개속도가 너무 빠르면 취출 상태가 변화하여 품질의 변동을 야기 할 수 있다.

4. Mold Clamping Force

The mold clamping force is the force for closing the parting surface, and its unit is kN (kilo Newtons) of tf (tons force). Although this unit is usually expressed merely as "tons" while speaking, in mechanics it is correct to append the weight unit "f".

4. 금형의 형체력

형체력은 파팅면을 닫기 위한 힘이며, 단위는 kN (kilo Newtons) tf (tons force) 이다. 이 단위는 말하자면 일반적으로 "톤"으로 표현되지만, 역학에는 중량 단위 "f"를 추가하는 것이 옳다.

5. Ejecting Speed

The ejecting speed is the speed of forward movement by the ejector rod, and its unit is ㎜/s. If the ejecting

5. 취출속도

취출속도는 취출로드에 의한 전진 속도이며 단위는 ㎜/s 이다.

speed is too high, the product may get deformed or the surface may become white. If the ejecting speed is too low, the molding cycle time becomes long.

취출속도가 너무 빠르면 제품이 변형되거나 표면이 하얗게 될 수 있다. 취출속도가 너무 낮으면 성형 사이클 시간이 길어진다.

6. Ejection Count

The ejection count or the number of times of ejection is very often set in the range of 1 to 3. Although ejecting once is enough if the mold is one with good mole release property; however, if it is bad, it may not be possible to take out the molded product from the mold unless ejection is made several times.

6. 취출 횟수

취출 횟수 또는 토탈 취출 회수는 1 내지 3의 범위로 설정되는 경우가 많다. 금형의 이형성이 양호한 경우에는 1 회 취출로도 충분하지만, 취출 불량의 경우, 여러 번 취출하지 않으면 금형으로부터 성형품을 빼내지 못할 수도 있다.

B 018 Releasing the Molded article from the Mold

In plastic injection molding, it is necessary to take out the molded article certainly from the mold.

Since the biggest advantage of injection molding is that it is possible to carry out mass production without requiring a human operator, if the molded article remains in the mold and it becomes necessary to take it out manually by a human operator, an operator will be required to be present at all times near the molding machine, automated operation cannot be made, and hence the personnel cost becomes very high.

If by chance, a mold is closed with a molded article still remaining in it, in the worst case, the cavity or the core will be damaged causing a big economic accident.

B-018 금형으로부터 성형품 취출하기

사출성형 시에 성형품을 금형으로부터 확실히 꺼낼 필요가 있다.

사출성형의 가장 큰 장점은 작업자 없이 대량 생산이 가능하다는 점에서 성형품이 금형 내에 남아 있고, 작업자가 수동으로 꺼내야 할 경우 작업자가 필요하다. 성형기 근처에 항상 상주해야 하기 때문에 자동 운전을 할 수 없어 인건비가 대단히 높아진다.

우연히 성형품이 남아있는 상태에서 금형을 닫으면 최악의 경우 캐비티 또는 코어가 손상되어 큰 경제적 사고를 초래한다. 금형이 손상된

If a mold is damaged, in most cases it is not possible to repair it easily, and hence it is necessary to adjust the mold so that the molded article can be taken out 100% from the mold.

Now, why does a molded article remain in the mold? A molded article remaining in the mold is called a mold release defect. A mold release defect occurs due to a combination of the following causes.

1) Contraction of the molded article
2) Wall thickness of the molded article
3) Internal diameter of the molded article
4) Ease of sliding of the plastic resin
5) Surface roughness of the cavity and core
6) The draft angle of the cavity and core
7) The surface temperature of the molded article at the time of opening the mold
8) Ejection speed
9) Number and placement of the ejection pins
10) Corner shape of the molded article
11) The state of vacuum between the molded article and the cavity as well as the core
12) Orientation of the glass fibers, etc.

Since it is very difficult to take countermeasures counter to common perception when a mold release defect has occurred, it is very important to give sufficient considerations at the time of the initial investigations during the design of a mold. In specific terms, the key point is to carry out sufficient investigations of the shape and material of the molded article, gate position, the draft angle, the height of the molded article, placement of the slide core or the inclined pin, the wall thickness of the molded article, etc.,

경우에는 대부분의 경우 쉽게 수리할 수 없기 때문에 성형품을 100% 빼낼 수 있도록 금형을 조정할 필요가 있다.

현재, 왜 성형 된 제품이 금형에 남아 있을까? 금형 내에 잔류하는 성형품을 이형 불량이라고 한다. 다음과 같은 원인의 조합으로 인하여 이형 불량이 발생한다.

1) 성형품의 수축
2) 성형품의 살 두께
3) 성형품의 내경
4) 플라스틱 수지의 슬라이딩 용이성
5) 캐비티와 코어의 표면 거칠기
6) 캐비티와 코어의 구배 각도
7) 금형 개방 시의 성형품 표면 온도
8) 취출 속도
9) 취출 핀의 수와 배치
10) 성형품의 코너 형상
11) 성형품과 캐비티 및 코어 사이의 진공 상태
12) 유리 섬유 등의 배향

이형 불량이 발생했을 때 일반적인 대책을 세우는 것은 매우 어렵기 때문에 금형설계 초기부터 충분한 고려가 필요하다. 구체적으로 성형품의 형상 및 재질, 게이트 위치, 구배 각도, 성형품의 높이, 슬라이드 코어 또는 경사 핀의 배치, 살 두께 등의 충분한 조사가 중요하다.

and how you can foresee the locations where a mold release defect can occur. Since it is not possible to take countermeasures if it is not possible to foresee such possible occurrences, it is very important to fully understand samples of various previous mold release defects and defect analysis reports.

In other words, it is very important that history reports have been accumulated thoroughly of all the mold release defects that have occurred in the past. Regarding defects that are caused by a combination of reasons such as mold release defects, it is difficult to find the causes from the analyses. Although a veteran designer can foresee such possible causes of mold release defects through tacit knowledge, this is very difficult if your experience is not sufficient. Because of this, it is critical to systematically accumulate data where previous occurrences of mold release defects are appropriately recorded.

이형 결함이 발생할 수 있는 위치를 예측할 수 있는 방법에 대해 설명한다. 이형 불량 가능성을 예측할 수 없다면 대응책을 취할 수 없기 때문에 이전의 다양한 이형 결함 및 결함 분석 보고서 샘플을 완전히 이해하는 것이 매우 중요하다.

다시 말해 과거에 발생한 모든 이형 결함에 대한 과거 보고서 축적이 매우 중요하다. 이형 불량 등의 원인이 복합적으로 생기는 것에 관해서는 원인 분석이 곤란하다. 베테랑 설계자는 암묵적인 지식을 통해 이형 불량의 원인을 예측할 수 있지만 경험이 충분하지 않은 경우 매우 어렵다. 이 때문에 이전에 금형 이형 불량이 적절하게 기록 된 데이터를 체계적으로 축적하는 것이 중요하다.

C. 성형방법의 종류

C 001 Molding Technology for Plastics with High Cavity Surface Temperatures

In engineering plastics resin and super engineering plastics resin, the types of plastics resin in which case the surface temperature of the cavity exceeds 100℃ is increasing. If the surface temperature of the cavity exceeds about 90℃, normally it becomes difficult to increase the temperature when the temperature control is being made using hot water.

In general, the following methods are used.

1. Temperature control using oil

In the temperature control using oil, oil ejected from a re-circulating pump is re-circulated via a joint hose through a flow path provided in the mold plate and cavity, thereby maintaining a constant temperature. Once the temperature rises up to the set temperature, it is possible to maintain a relatively stable temperature. However, there is the disadvantage that it takes time for the temperature to rise. Further, there is the danger of people suffering burns while handing the oil, and also there is the problem that later treatment of the oil is cumbersome.

2. Electric heater

In the temperature control using an electric heater (cartridge heater), the temperature is maintained constant using the heater in combination with a temperature sensor (thermocouple, etc.). Since the thermal capacity is high,

C-001 고온의 캐비티가 필요한 플라스틱의 성형 기술

엔지니어링 플라스틱 수지 및 슈퍼 엔지니어링 플라스틱 수지의 경우 캐비티의 표면 온도가 100°C를 초과하는 플라스틱 수지의 유형이 증가하고 있다. 캐비티의 표면 온도가 약 90°C를 초과하면 일반적으로 뜨거운 물을 사용한 온도 제어방법으로는 온도를 높이기가 어려워진다.

일반적으로 다음과 같은 방법이 사용된다.

1. 기름을 이용한 온도 조절

기름을 이용한 온도 제어에 있어서 재순환 펌프로부터 토출 된 기름은 형판과 캐비티에 설치된 유로를 통과하여 조인트 호스를 통해 재순환되어 온도가 일정하게 유지된다. 온도가 설정 온도까지 상승하면 비교적 안정된 온도를 유지할 수 있다. 그러나 온도가 상승하는 데에는 시간이 걸리는 단점이 있다. 또한, 기름을 작업하는 동안 사람들이 화상을 입을 수 있는 위험이 있으며, 작업이후 기름의 처리가 번거롭다는 문제가 있다.

2. 전기 히터

전기 히터 (카트리지 히터)를 사용하는 온도 제어에는 온도 센서 (써버 커플 등)와 함께 히터를 사용하여 온도를 일정하게 유지한다. 열

there is the advantage that the temperature rise is fast.

However, the temperature near the heater is high and the temperature far from the heater is low and hence it becomes difficult to maintain a uniform temperature distribution.

In addition, since a heater has a limited life, it will be necessary to replace the heater periodically.

Mounting holes and clearances will be very important in mounting the heater. If the clearance is too much, the heater will be in the no load condition and hence its life will be shortened.

In order to carry out more accurate temperature control, it is recommended to place a heat insulating plate between the platen of the injection molding machine and the mold installation plate. Or else, it is also possible to provide heat insulating plates surrounding the mold.

용량이 높기 때문에 온도 상승이 빠르다는 장점이 있다.

그러나 히터 근처의 온도가 높고 히터로부터 멀리 떨어지면 온도가 낮아 균일한 온도 분포를 유지하기가 어려워진다.

히터의 수명이 짧기 때문에 정기적으로 히터를 교체해야한다.

히터를 장착 할 때 설치할 홀과의 공차가 매우 중요하다. 클리어런스가 너무 크면 히터가 무부하 상태가되므로 수명이 단축된다.

보다 정확한 온도 제어를 수행하려면 형판 사이에 단열재를 배치하는 것이 좋다. 사출기 및 금형 설치판 또는 금형을 둘러싸는 단열판을 설치하는 것도 가능하다.

Plastic name	Cavity surface temperature (℃)	Plastic name	Cavity surface temperature (℃)
PPS (Polyphenylene sulfide) with 30% glass fiber	130 to 150	PC (Polycarbonate)	80 to 120
LCP (Liquid crystal polymer) with 40% glass fiber	70 to 110	Heat resistant PLA (polylactic acid)	110 to 120
PET (Polyethylene terephthalate)	130 to 150	PEEK (Polyether ether ketone)	120 to 160
PA46 (Polyamide/Nylon 46)	80 to 120	PI (Polyimide)	170 to 200

Table 1 Example of materials requiring high cavity surface temperatures

In the case of molds for the injection molding of PPS or liquid crystal polymer, since it is necessary to maintain the cavity surface temperature at more than 100℃, temperature control using oil or temperature control using a cartridge heater will be required. The temperature controlling method used very frequently with the cartridge heaters available in the market is the "ON-OFF Control" method in which the temperature is controlled by turning ON and OFF the power supply to the heater.

Since this "ON-OFF Control" can be constructed with a simple switching device, although the price of the controller is relatively low, the variations in the cavity surface temperature are large, and there is the drawback that it is difficult to stabilize the temperature.

If the cavity surface temperature is unstable, it will cause fluctuations in the condition of shrinkage, the dimensions, and the glossiness of the surface in a precision molded item.

In view of this, the "PID Control" method is recommended for stabilizing the cavity surface temperature as much as possible. The PID Control is a method of controlling that can shorten the time required for the temperature to become stable by using controls of the following types:

- Proportional - Integral - Derivative.

Cartridge heaters can have a maximum load of 10 kW with an AC three phase 200 V power supply.

Therefore, as an example of use, it is possible to use two three phase 200 V 1 kW heaters and to incorporate them at the movable side mold plate and the fixed side mold plate, and to carry out stable control of the cavity surface temperature using PID control.

PPS 또는 액정 폴리머의 사출성형을 위한 금형의 경우 캐비티 표면 온도를 100℃ 이상으로 유지해야하기 때문에 기름을 사용하는 온도 제어 또는 카트리지 히터를 사용하는 온도 제어가 필요하다. 시중에서 구입할 수 있는 카트리지 히터로 자주 사용되는 온도 제어 방법은 히터의 전원을 ON/OFF하여 온도를 제어하는 "ON-OFF 제어" 방식이다.

"ON-OFF 제어"는 간단한 스위칭 장치로 구성 할 수 있기 때문에 컨트롤러의 가격은 비교적 저렴하지만 캐비티 표면 온도의 변화는 크며, 온도를 안정화시키는 것이 어렵다는 단점이 있다.

캐비티 표면 온도가 불안정하면 정밀 성형품의 표면 수축 상태, 치수 및 표면 광택이 변동하게 된다.

이를 고려하여 캐비티 표면 온도를 가능한 한 안정화시키기 위해 "PID 제어" 방법을 권장한다. PID 제어는 다음과 같은 제어를 사용하여 온도가 안정화되는 데 필요한 시간을 단축 할 수 있는 제어 방법이다. :

-Proportional -Integral - Derivative.

카트리지 히터는 AC 3상 200V 전원 공급 장치를 사용하여 최대 10kW의 부하를 견딜 수 있다.

따라서 사용의 예로서, 2개의 3상 200 V 1 kW 히터를 사용하고, 가동측 금형과 고정측 금형에 이를 통합 할 수 있으며, PID 제어를 이용하여 캐비티 표면 온도를 안정적으로 제어 할 수 있다.

Heat insulating plates are very important parts for stabilizing the temperature of a mold for injection molding, or for realizing energy savings while maintaining the temperature. They are used as mandatory items in the injection molding of engineering plastics or super engineering plastics.
In general, the method of using a heat insulating plate can be of the following two types.

단열판은 금형의 온도를 안정화 시키거나 온도를 유지하면서 에너지 절약을 실현하기 위해 매우 중요한 부품이다. 엔지니어링 플라스틱 또는 수퍼 엔지니어링 플라스틱의 성형에서 필수적으로 사용된다.

일반적으로 단열재를 사용하는 방법은 다음 두 가지 유형이 될 수 있다.

1) Using by fixing to the platen of the injection molding machine.
2) Using by fixing behind the mounting plate of the mold.

1) 사출성형기의 고정반에 설치하여 사용
2) 금형의 설치판 뒤쪽에 설치하여 사용

1. Withstanding temperature

As a guideline, the following recommended usage temperatures are available.

- Recommended usage temperature:

Up to 100℃, Up to 180℃, Up to 220℃, Up to 400℃, Up to 500℃.

1. 내열 온도

가이드라인으로서 다음과 같은 권장 사용 온도가 있다.

- 권장 사용 온도 :

최대 100°C, 최대 180°C, 최대 220°C, 최대 400°C, 최대 500°C.

2. Material

The material is the factor related to the recommended usage temperature and the compression strength.

The following materials are available.

- Cotton cloth + phenol resin (Bakelite)
- Craft paper + phenol resin (Bakelite)
- Glass fiber + silicate binder
- Glass fiber + ultra high heat resistant epoxy resin

2. 재질

재료는 권장 사용 온도 및 압축 강도와 관련된 요소이다.

다음 재료를 사용할 수 있다.

- 면직물 + 페놀 수지 (베이클라이트)
- 공예 용지 + 페놀 수지 (베이클라이트)
- 유리 섬유 + 규산염 결합제
- 유리 섬유 + 초고 내열성 에폭시 수지

- Glass fiber + phosphate binder
- Glass fiber + borate binder

- 유리 섬유 + 인산염 결합제
- 유리 섬유 + 붕산염 결합제

C 002 Special Injection Molding Methods (Two Color Injection Molding Method)

Special methods of plastic injection molding are explained starting from this lesson. Firstly, in this lesson we describe the two color injection molding method.

The two color injection molding method is a method of molding that is recently being called the "two material injection molding method" or the "different material injection molding method", etc. Two types of thermoplastic resin are successively injected into the mold by respectively different injection cylinders, thereby producing a product with two types of colors.

This is a molding technology that is being used widely in the preparation of key tops for high end desktop PCs, or the illuminated buttons of car navigation units, etc.

In general, it appears that very often two plastic resin of the same type such as PS plastic or ABS plastic are used. This is because there is very good adhesion between the two molded items. Although it is possible to produce molded products from two different types of plastics resin such as ABS and POM, the adhesion between them is not necessarily good. (There are different applications when the adhesion is good and when the adhesion is not good.)

C-002 특수 사출성형법 (2색 사출성형법)

이번 강의를 시작으로 특수 플라스틱 사출성형 방법을 설명한다. 첫째, 이 강의에서는 2색 사출성형 방법을 설명한다.

2색 사출성형법은 최근에 "두 가지 재료 사출성형법" 또는 "이종 재료 사출성형법"등으로 불리는 성형 방법이다. 두 종류의 열가소성 수지를 각각 다른 실린더를 이용하여 사출성형한다. 따라서 두 가지 색을 갖는 제품이 성형된다.

이것은 고급사양의 데스크탑 PC용 키탑 또는 자동차 네비게이션 장치의 조명 버튼 등을 성형하는데 널리 사용되는 성형기술이다.

일반적으로 PS 플라스틱 또는 ABS 플라스틱 처럼 같은 유형의 두 가지 플라스틱이 사용된다. 왜냐하면 성형품 사이의 접착력이 매우 좋기 때문이다. ABS 및 POM과 같은 두 가지 다른 유형의 플라스틱 수지로 성형품을 생산할 수는 있지만, 이들 사이의 접착력은 반드시 좋지는 않다.(양호한 접착력이 필요하거나 그렇지 않은 경우에 따라 적용이 다르다.)

[Fig. 1] Example of two color molding method

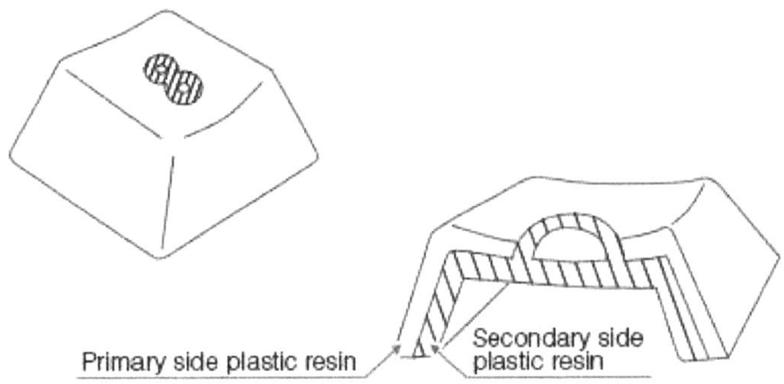

In addition, recently there are some unique combinations that have been realized such as combinations of thermoplastic plastic material with a thermoplastic elastomer (rubber-like plastic resin). (Sports goods, etc.)

In order to practically use the two color injection molding method, usually, a special injection molding machine will be necessary. Such machines are being made by manufacturers of injection molding machines in Japan as well as in countries like Switzerland and Germany. The injection molding machine is equipped with two injection units, which respectively pour the molten material into the interior of the cavity of the mold via their respective sprues.

In the mold, the female part of the cavity is formed on the fixed side of the respective plastic material.

On the other hand, two male cores of the same shape are formed on the moving half, and the space between the male parts can be moved by a rotating mechanism or a sliding mechanism. (There are several types of patterns of this structure.)

또한, 최근에는 열가소성 플라스틱 재료와 열가소성 엘라스토머(고무상 플라스틱 수지)의 조합과 같이 특별한 경우도 있다. (스포츠 용품 등)

실제로 2색 사출성형법을 사용하기 위해서 특수 사출성형기가 필요하다. 이러한 사출성형기는 일본의 사출성형기 제조업체 및 스위스 및 독일과 같은 국가에서 생산되고 있다. 사출성형기에는 두 개의 사출장치가 장착되어 있으며, 각각 사출장치가 용융된 재료를 각각의 스프루를 통해 금형의 캐비티 내부로 사출한다.

금형에서 캐비티는 고정측 금형에 형성된다.

다른 한편으로는 동일한 형상의 2개의 코어가 이동측 금형에 형성되고, 고정측 금형은 회전기구 또는 슬라이딩 기구에 의해 이동 될 수 있다. (이 구조에는 몇 가지 유형의 패턴이 있다.)

In the two color injection molding method, since a beautiful multi-function molded product can be produced in one step, it is possible to produce a molded item with high value addition. It is also possible to have multiple cavities in a single shot in the case of molded items with small sizes.

However, the design of the molds requires knowledge about the design of wall thicknesses and know-how about the bonding between different plastic materials. Some techniques will be necessary regarding the temperature control of molds also.

2색 사출성형법에서는 한 번에 아름다운 다기능 성형품을 제조 할 수 있기 때문에 부가 가치가 높은 성형품을 성형 할 수 있다. 작은 크기의 성형품인 경우에는 한번의 성형으로 여러 캐비티를 성형 할 수 있다.

그러나 금형설계에는 다양한 플라스틱 재료 간의 결합에 대한 살두께 및 노하우 설계에 대한 지식이 필요하다. 금형의 온도 제어와 관련하여 몇 가지 기술이 필요하다.

[Fig. 2] Example of two color molding method

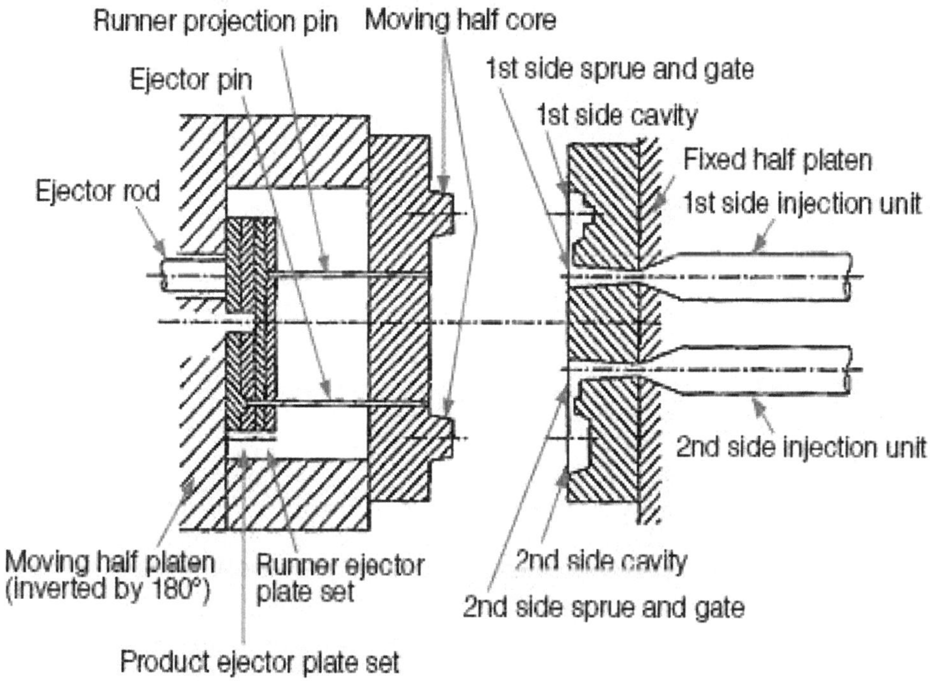

C 003 Special Injection Molding Methods(Blow Molding Method)

Blow molding is a method of injection molding hollow molded items such as PET (polyethylene terephthalate) bottles for juice, etc. This method is being used widely in the world for manufacturing shampoo containers, containers for spices such as soy sauce, or containers for detergents, etc. The plastic resin that can be blow molded are polyethylene terephthalate, polypropylene, polyvinyl chloride, nylon, polycarbonate, etc.

Blow molding is done using a special blow molding machine. And also, the mold used has only a female type cavity and there is no male type core. Instead of that, there is a nozzle for blowing air into the mold, a pre-formed balloon shaped molded item called a parison is blown and made to contact the cavity closely thereby transferring the shape.

In the case of blow molding, usually a relatively soft metal is used for the cavity. While aluminum alloys, bronze, etc. are commonly used, special steels are also used.

Since very frequently beautiful glossiness is expected in the surface of the molded item, sufficient care has to be taken about polishing the inside surface of the cavity.

Further, multiple cavities in a mold are common in the blow molding of containers for drinks, and since a high cycle rate is expected, techniques will be necessary for the function of cooling and adjusting the temperature of the mold.

C-003 특수 사출성형법 (블로우 성형법)

블로우 성형은 주스용 PET(폴리에틸렌 테레프탈레이트)병과 같은 중공 성형품을 사출성형하는 방법이다. 이 방법은 샴푸 용기, 간장과 같은 향신료용 용기 또는 세제용 용기 생산용으로 세계에서 널리 사용되고 있다. 블로우 성형이 가능한 플라스틱 수지로는 폴리에틸렌 테레프탈레이트, 폴리프로필렌, 폴리염화비닐, 나일론, 폴리카보네이트 등이 있다.

블로우 성형은 특수 블로우 성형기를 사용하여 수행된다. 또한 사용된 금형에는 캐비티만 있고, 코어는 없다. 그 대신 금형에 공기를 불어넣기 위한 노즐이 있으며 파리손이라고 불리는 예비 성형 된 벌룬 형상의 성형품을 블로우 성형하여 캐비티에 밀착시켜 형상을 전사한다.

블로우 성형의 경우 통상 비교적 부드러운 금속이 캐비티에 사용된다. 알루미늄 합금, 청동 등이 일반적으로 사용되지만 특별한 강재도 사용된다.

성형품 표면에 매우 아름다운 광택이 예상되므로 캐비티의 내부 표면을 연마하는 데 충분한 주의를 기울여야한다.

또한, 일반적으로 금형내의 다수 캐비티는 음료용 용기의 블로우 성형에 적용되고, 높은 사이클 률이 기대되기 때문에 냉각 및 금형 온도의 조정 기능에 대한 기술이 필요하다.

Since most of the blow molded items are of the disposable variety, it is extremely important that the product design and planning and the mold design are made considering the recycling of these products.

블로우 성형품의 대부분은 일회용품이므로 제품의 설계, 기획 및 금형설계는 이들 제품의 재활용을 고려하여 매우 중요하다.

C 004 Special Injection Molding Methods (Insert Method)

C-004 특수 사출성형 방법 (인서트 성형법)

Insert molding is a method of injection molding in which an insert (a metallic part or other inserted item) is installed inside the cavity while the mold is still open, and the mold is then closed and the injection molding is carried out. This method is used for the injection molding of housings in which brass screws or shafts are insert molded, or for electronic components in which metal terminals or contacts are insert molded.

The simplest method of placing the inserts inside the cavity is to affix to the cavity by hand or by robots the inserts which are machined separately as individual parts (Fig. 1).

인서트 성형은 금형이 열려있는 동안 캐비티 내부에 인서트 (금속 부품 또는 기타 삽입물)를 삽입한 다음 금형을 닫고 사출성형하는 방법이다. 이 방법은 황동 나사 또는 축(軸)이 인서트 성형된 하우징의 사출성형 또는 금속 단자 또는 접점이 인서트 성형된 전자 부품용의 성형에 적용된다.

캐비티 내부에 인서트를 삽입하는 가장 간단한 방법은 개별 부품으로 개별적으로 가공 된 인서트 부품을 손 또는 로보트로 캐비티에 삽입한다. (그림 1).

[Fig. 1] Example of insert molded items

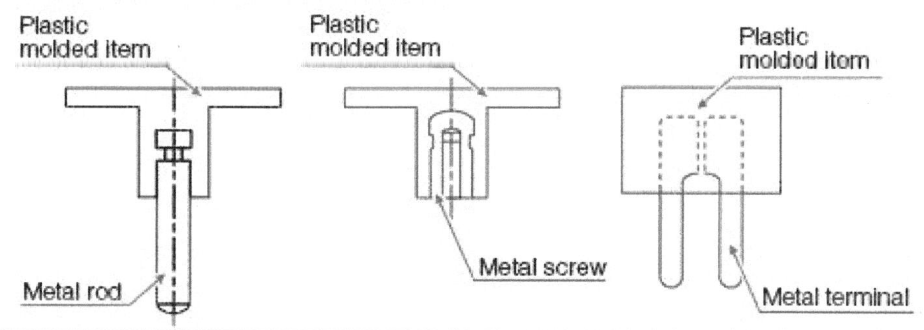

[Fig. 2] Hoop feeding type insert molding method

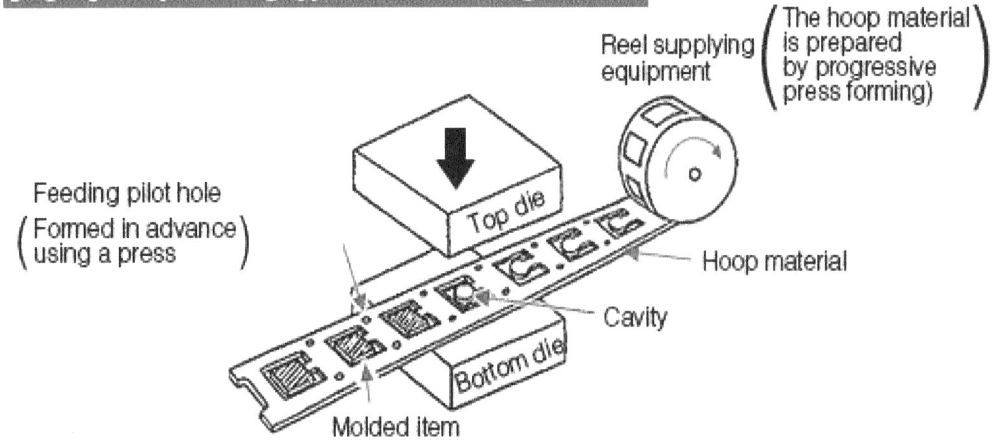

In order to manufacture a large quantity of a part efficiently, there has also been proposed a method of progressive molding in which the molding is done successively after automatically feeding into the cavity a metal hoop in which the required shapes have been formed previously by a progressive press forming die (Fig. 2).

Since the insert molding method is a method for manufacturing a molded item by a combination of a metal formed part and plastic, it is necessary to accumulate know-how by making trial and error in order to combine material with different properties such as thermal expansion coefficient, heat transfer coefficient, strength, etc.

Good injection mold design will be difficult if the accuracies and capacities of press forming are also not understood.

Insert molding is a method of molding that offers a high value addition in addition to making it possible to manufacture at a low cost various precision electronic components or excellent quality molded items for automobile parts.

다량의 부품을 효율적으로 생산하기 위해서 프로그래시브 프레스 성형 다이에 의해 소정의 형상이 미리 형성되어있는 금속 후프를 캐비티에 자동적으로 공급 한 후에 연속적으로 성형을 행하는 프로그래시브 성형 방법이 제안되어 있다(그림 2).

인서트 성형법은 금속 성형품과 플라스틱을 조합하여 성형품을 제조하는 방법이기 때문에 열팽창 계수, 열전달 계수, 강도 등의 성질이 다른 재료를 조합하기 위해서 시행착오를 거쳐 노하우를 축적 할 필요가 있다.

프레스 성형의 정확성과 용량을 이해하지 못하면 좋은 금형설계가 어려울 수 있다.

인서트 성형은 저가의 다양한 정밀 전자 부품 또는 우수한 품질의 자동차 부품을 제조 할 수 있을 뿐만 아니라 부가가치가 높은 성형 방법이다.

On the contrary, without a high level of technical ability, it is not possible to manufacture large quantities in a stable manner. The precision insert molding technology and mold preparation technology of Japan can be said to be of the top level in the world.

In order to realize insert molding practically, the excellence in the press forming die design and fabrication technology, for example, the technology of positioning or the technology of mold part machining, etc., need to be incorporated into the plastic mold design technology in a flexible manner.

반대로 고도의 기술력이 없으면 안정적으로 대량 생산이 불가능하다. 일본의 정밀 인서트 성형 기술 및 금형 제작 기술은 세계 최고 수준이라고 할 수 있다.

실제로 인서트 성형을 실현하기 위해서 프레스 성형 금형설계 기술의 우수성, 예를 들면 위치 결정 기술이나 금형 부품 가공 기술 등을 유연한 방식으로 플라스틱 금형설계 기술에 통합 할 필요가 있다.

[Fig. 3] Example of insert molded item

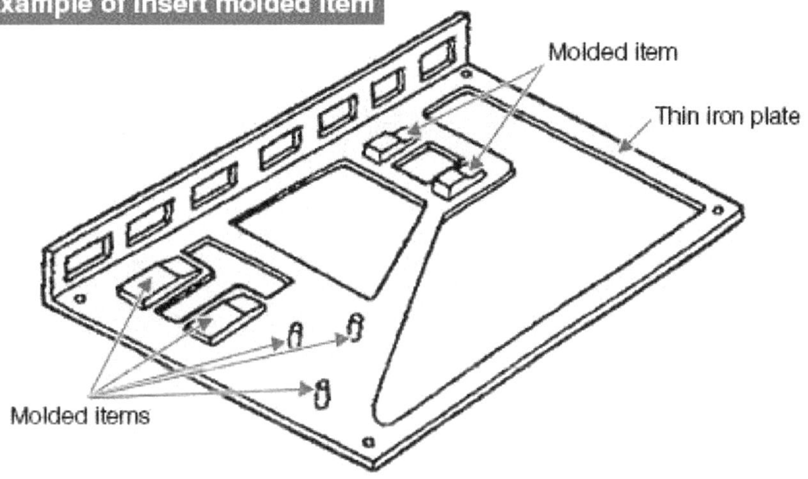

C 005 Compression Molding Method

C-005 압축 성형법

Compression molding is a method of molding used for molding thermoset plastic resin. Thermoset plastic resin are liquid at room temperature, but solidify upon

압축 성형은 열경화성 플라스틱 수지 성형에 사용되는 성형 방법이다. 열경화성 플라스틱 수지는 실온에서 액체이지만 가열되면 고화되

heating, and once solidified, they remain permanently solid. The hardening is considered to be due to a chemical reaction called cross-linking.

고, 일단 고화되면 영구적으로 유지한다. 경화는 가교 결합이라고 하는 화학 반응으로 인한 것이다.

Typical thermoset plastic resin are phenol resin, melamine resin, urea resin, epoxy resin, etc.

일반적인 열경화성 플라스틱 수지는 페놀 수지, 멜라민 수지, 우레아 수지, 에폭시 수지 등이다.

The compression molding method is a classical method of molding which has been in use ever since plastic resins were invented.

압축 성형 방법은 플라스틱 수지가 발명된 이래로 사용 되어온 고전적인 성형 방법이다.

Male and female molds are prepared, the molding resin is loaded inside the female mold, the male mold is closed, and the set of molds is heated thereby solidifying the plastic resin and preparing the molded article.

캐비티와 코어 금형을 준비하고, 캐비티 금형 내에 수지를 넣고, 코어 금형을 닫아 금형을 가열하여 플라스틱 수지를 고화시켜 성형품을 성형한다.

At the time of closing the male mold, excess resin overflows from the parting surface and becomes burrs. These burrs are removed after molding. (Fig. 1)

코어 금형을 닫을 때 과도한 양의 수지가 파팅면으로 넘쳐 버린다. 이러한 버는 성형 후에 제거된다. (그림 1)

[Fig. 1] Example of molds for compression molding (flash mold)

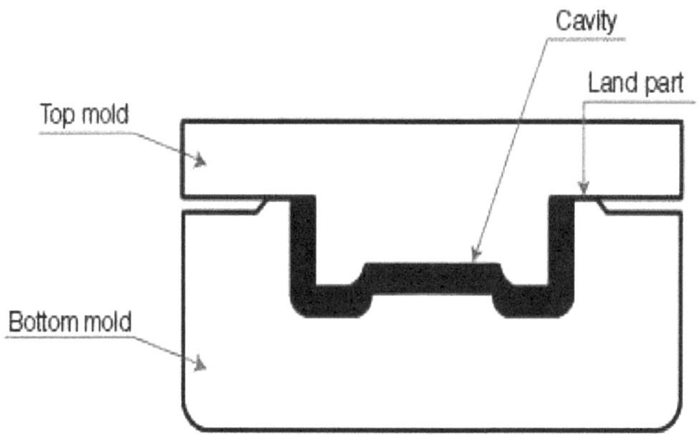

Compression molding is being used even now for food utensils such as plates, bowls, etc., electrical insulating parts, decorative buttons, etc.

접시, 그릇 등의 식품기구, 전기 절연 부품, 장식용 버튼 등에도 압축 성형이 사용되고 있다.

Since thermoplastic materials cannot be used in environments where they are heated or where resistance to chemicals is required, thermoset plastic resin are used in such situations.

열가소성 재료는 가열되거나 화학 물질에 대한 내성이 요구되는 환경에는 사용할 수 없으므로 열경화성 플라스틱 수지가 이러한 상황에서 사용된다.

The mold structures of the three types:

1) Positive type mold
2) Flash type mold
3) Semi-positive type mold.

세 가지 유형의 금형 구조 :

1) 포지티브 형 금형
2) 플래쉬 타입 금형
3) 세미 포지티브 타입 금형

The semi-positive type mold is used most often.

세미 포지티브 타입 금형이 가장 자주 사용된다.

C 006 Transfer Molding Method

C-006 트랜스퍼 성형법

The transfer molding method is a molding method of thermoset plastic resin. Although the compression molding method described last time is a typical molding method of thermoset plastic resin, the applied pressure may become non-uniform when the mold is closed and the pressure is applied, and as a result, it is possible that thin pins or some parts of the mold get deformed and damaged.

The transfer molding is a molding method that has been proposed as an improvement that corrects such weaknesses in the compression molding method.

The injection of the plastic to the interior of the mold is carried out by a plunger. Therefore, it is possible to inject the plastic to the interior of the mold with a relatively uniform pressure.

트랜스퍼 성형법은 열경화성 플라스틱 수지의 성형법이다. 마지막으로 설명된 압축 성형 방법은 열경화성 플라스틱 수지의 전형적인 성형 방법이지만, 금형이 닫히고 압력이 가해질 때 가해지는 압력이 불균일해질 수 있으며, 결과적으로 얇은 핀 또는 금형의 일부가 변형되어 손상된다.

트랜스퍼 성형은 압축 성형법에서 이러한 약점을 개선하는 개선책으로서 제안된 성형 방법이다.

금형 내부로의 플라스틱 주입은 플런저(plunger)에 의해 수행된다. 따라서 비교적 균일한 압력으로 금형의 내부에 플라스틱을 주입 할 수 있다.

The molds for this method can be broadly classified into the following two types.

이 방법의 금형은 다음 두 가지 유형으로 크게 분류 할 수 있다.

- Hot type molds
- Plunger type molds

- 핫 타입 금형
- 플런저 타입 금형

The transfer molding method is used very frequently for molding the sealing of semiconductor MPUs, or for molding electronic components (device components).

Measures for preventing inclusion of gases or bubbles that are generated during molding and the balancing of the runners in multiple cavity molds are important factors of know-how in this method of molding.

트랜스퍼 성형법은 반도체 MPU의 밀봉 또는 전자 부품 (디바이스 부품)의 성형에 매우 빈번하게 사용된다.

성형 중에 생성되는 가스 또는 기포의 포함을 방지하기 위한 조치 및 다중 캐비티 몰드에서의 런너 밸런싱은 이 성형법에서 중요한 노하우이다.

C 007 Vacuum Molding Method, Compressed Air Molding Method

C-007 진공 성형법, 압축 공기 성형법

The vacuum molding method is a method used for molding articles such as egg packages, pudding cups, packaging containers, etc. that are made of thin plastic sheets.

In the vacuum molding method, melting of raw material pellets such as in extrusion molding is not made, but a plastic plate formed in advance in the shape of a sheet is deformed by vacuum suction on to a heated mold.

진공 성형법은 얇은 플라스틱 시트로 만들어진 계란 포장, 푸딩 컵, 포장 용기 등과 같은 물품을 성형하는 데 사용되는 방법이다.

진공 성형법에서는 압출 성형에서와 같이 원료 펠릿의 용융은 이루어지지 않지만, 미리 시트 형상으로 성형한 플라스틱 판을 가열 된 금형에 진공 흡입하여 변형시킨다.

[Fig.] Outline of vacuum molding method

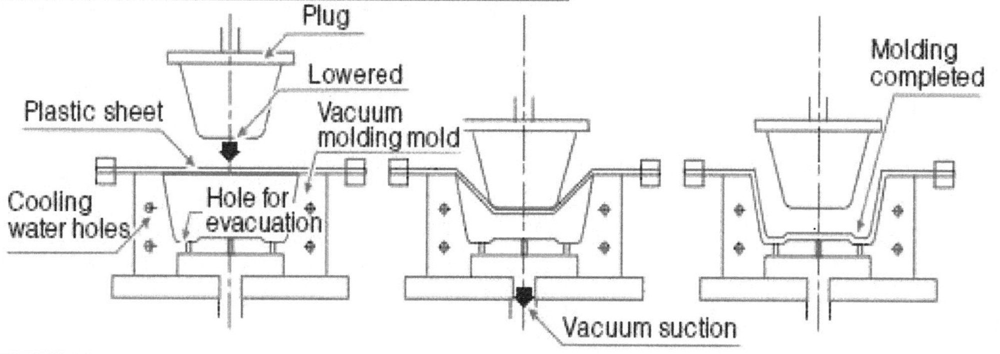

The mold is provided with several holes for vacuum suction. In order to support the deformation of the plastic sheet, a plug is also sometimes lowered to press down the sheet.

On the other hand, the compressed air molding is a molding method in which the plastic sheet is deformed by compressed air.

The edges of the items molded using these methods are cut by a press in a succeeding step or else, they are cut by scissors if such parts are small in quantity.

The plastics that can be molded in this manner are:

- Polystyrene foam
- Hard PVC
- Polypropylene
- Polyethylene terephthalate
- Methyl ester methacrylate (acrylic plastic), etc.

It is only necessary to prepare either the male mold or the female mold.

The materials used for the mold are mainly nonferrous metals such as an aluminum alloy, duralumin, zinc alloy, brass, etc.

금형에 진공 흡입을 위한 여러 구멍이 있다. 플라스틱 시트의 변형을 지지하기 위해 때로는 플러그가 하강되어 시트를 누른다. 한편, 압축 공기 성형은 플라스틱 시트가 압축 공기에 의해 변형되는 성형 방법이다.

이러한 방법으로 성형된 제품의 가장자리는 후속 단계에서 프레스로 절단되거나 그렇지 않으면 가위로 절단된다.

이 방법으로 성형 할 수 있는 플라스틱 재료는 다음과 같다.

- 폴리스티렌 폼
- 하드 PVC
- 폴리프로필렌
- 폴리에틸렌 테레프탈레이트
- 메틸 에스터 메틸에스테르(아크릴 플라스틱) 등

코어 또는 캐비티 중 하나만 준비하면 된다.

금형에 사용되는 재료는 주로 알루미늄 합금, 두랄루민, 아연 합금, 황동 등의 비철 금속이다.

C 008 Extrusion Molding

C-008 압출 성형

Extrusion molding is a method of continuously molding plastic molded articles having the same cross-sectional shape over its length such as rainwater guttering, gray colored drain water pipe, etc.

A dedicated extrusion molding machine is required for carrying out extrusion molding. In addition, dies are also required for obtaining the shape of the molded item. A die for extrusion molding is generally called an extrusion die.

There are various types of extrusion dies such as the following.

- Straight die
- Cross-head die
- Flat die
- Die for solids
- Die for films
- Die for pipes
- Die for coverings

Extrusion dies are prepared from stainless steel, nickel-chromium steel, or alloy tool steel. The dies are hard-chrome plated or surface treated so as to withstand corrosion and wear.

The holder part of the die is made of a soft steel such as S50C, etc.

The shape and design of the die are very important factors determining the quality of the molded item. The processing conditions such as the speed of extrusion are also important factors.

압출 성형이란 빗물받이, 회백색 배수관 등의 단면형상이 동일한 플라스틱 성형품을 연속 성형하는 방법이다.

압출 성형을 수행하기 위해서 전용 압출 성형기가 필요하다. 또한, 성형품의 형상을 얻기 위해서 다이가 필요하다. 압출 성형용 금형은 일반적으로 압출 다이라고 한다.

다음과 같은 다양한 유형의 압출 다이가 있다.

- 스트레이트 다이
- 크로스 헤드 다이
- 평 다이
- 고체용 다이
- 필름용 다이
- 파이프용 다이
- 커버용 다이

압출 다이는 스테인리스 강, 니켈-크롬 강 또는 합금 공구강으로 제작된다. 다이는 경질 크롬 도금 또는 표면 처리되어 부식과 마모를 견딜 수 있다.

다이의 홀더 부는 S50C 등의 연강으로 구성되어 있다.

다이의 모양과 디자인은 성형품의 품질을 결정하는 매우 중요한 요소이다. 압출 속도와 같은 가공 조건 또한 중요한 요소이다.

Extrusion molding is the second most frequently used method of molding plastic molded articles next only to injection molding. Unlike injection molding, once a die is prepared, a relatively large quantity can be produced from the die. Therefore, this can be said to be a molding method that finds a lot of applications in the equipment manufacturing industry.

압출 성형은 플라스틱 성형품을 사출성형으로 만드는 것 다음으로 빈번하게 사용되는 방법이다. 사출성형과 달리, 일단 다이가 준비되면 다이로부터 비교적 많은 양이 생산될 수 있다. 따라서 압출 성형은 제조업 분야에 많이 응용되는 성형방법이라 할 수 있다.

CHAPTER D

D. 성형불량과 대책

D 001 Solutions for Molding Defects (Sink Marks)

Sink marks arise from the phenomenon when the surface of the molded product shrinks and becomes slightly depressed.

This may cause quality defects in the case of molded products whose outer surface is important as a product.

The following measures can be taken to solve the problem of sink marks.

1. Measures related to molds

1) Slightly decreasing the cavity surface temperature.
2) Making the gate wider.
3) Making the runner size larger.
4) Making the sprue thicker.
5) Re-studying the cooling path of the mold, and increasing the cooling efficiency.
6) Changing the cooling structure of hard-to-cool parts to easy-to-cool parts. (Example: Baffle plate structure, cooling pipe structure, heat pipe, nests of non-ferrous metals)
7) Increasing the number of gates.
8) Changing the gate position to a thick wall part.

D-001 성형 결함에 대한 해결책 (싱크 마크)

싱크 마크는 성형품의 표면이 수축하여 약간 움푹 들어가게 되는 현상이다.

이는 외부 표면이 중요한 성형품의 경우에서 품질 결함의 원인이 될 수 있다.

싱크 마크 문제를 해결하기 위해 다음과 같은 조치를 취할 수 있다.

1. 금형 관련 조치사항

1) 캐비티 표면 온도를 약간 낮춤
2) 게이트 넓히기
3) 런너 크기를 증가시킴
4) 스프루를 굵게 하기
5) 금형의 냉각라인을 다시 연구해서 냉각 효율을 증가시킴.
6) 냉각하기 어려운 부품의 구조를 냉각하기 쉬운 구조로 변경 (예 : 배플 플레이트 구조, 냉각 파이프 구조, 히트 파이프, 비철금속 등)
7) 게이트의 수를 증가시킴
8) 두꺼운 살 두께 부분으로 게이트 위치를 변경하기

2. Measures related to injection molding conditions

1) Making the dwelling time longer.
2) Setting the dwelling pressure to a higher value.
3) Setting the injection speed to a higher value.
4) Lowering the nozzle temperature.
5) Try increasing the value of the measure.
6) Try increasing the weight of the cushion.
7) Try changing the injection molding machine.
8) Replacing the reverse flow prevention ring of the injection unit.

2. 사출성형 조건에 관한 조치사항

1) 보압 유지시간을 길게 하기
2) 보압을 더 높게 설정
3) 사출속도를 더 빠르게 설정
4) 노즐온도를 낮게 설정
5) 계량 값을 증가시킴
6) 쿠션량의 증가
7) 사출성형기의 변경
8) 사출장치의 역류 방지 링의 교체

3. Measures related to the design of the molded product

1) Removing the thick wall parts in the molded product.
(Example: Forming a recessed shape, using parts for other purposes)
2) Using a non-crystalline plastic

3. 성형품의 설계에 관한 조치사항

1) 성형품의 두꺼운 살 부분을 제거
(예 : 오목한 모양 만들기, 다른 목적으로 파트 사용)
2) 비결정성 플라스틱 사용

D 002 Solutions for Molding Defects (Short Shots)

D-002 성형 결함에 대한 해결책 (미성형)

A short shot is a phenomenon in which there is an incomplete filling in a part of the molded product.

There are two types of causes for short shots in terms of their nature. One of these occurs because, in the middle of the flow of the molten plastic, the front end of the flow gets cooled and solidifies. The second is

미성형은 성형품의 일부가 완전히 충전되지 못하는 현상이다.

미성형의 원인은 두 가지로 나눌 수 있다. 첫 번째 원인은 용융 된 플라스틱 수지가 유동하는 중에 유동선단이 냉각되어 고화되기 때문이다.

caused, in the flow process of the molten plastic, because air traps are generated in the flow depending on the conditions of the flow.

In order to take countermeasures against short shots, it is necessary to verify which of the above two types the short shot belongs to.

두 번째 원인은 용융수지가 유동하는 과정에서 유동 조건에 따라 에어 트랩이 발생하기 때문이다.

미성형에 대한 대책을 마련하기 위해서 미성형의 발생 원인이 두 가지 유형 중 어느 것에 해당하는지 확인하여야 한다.

1. Countermeasures related to molds

1) Widen the gate
2) Widen the runner.
3) Make the sprue thicker.
4) The cold slag well is small.
5) Provide a heat insulating plate on the under surface of the mold plate.
6) Increase the number of gates.
7) Change the position of the gates.

1. 금형 관련 대책

1) 게이트의 확대
2) 런너의 확대
3) 스프루의 확대
4) 콜드 슬러그 웰이 작다.
5) 형판 아래 표면에 절연 플레이트를 삽입
6) 게이트 수의 확대
7) 게이트의 위치 변경

2. Countermeasures related to the injection molding conditions

1) Make the plastic temperature higher.
2) Set the cavity surface temperature a little higher.
3) Try making the filling pressure higher.
4) Try making the dwell pressure higher.
5) Try setting the dwell pressure time longer.
6) Try increasing the charge quantity.
7) Try increasing the volume of the cushion.
8) Try changing the injection molding machine.
9) Replace the reverse flow prevention ring of the injection unit.
10) Change the tip diameter of the injection nozzle of the injection molding machine to a larger size.

2. 사출성형 조건에 관한 대책

1) 용융수지 온도를 더 높게 설정한다.
2) 캐비티 표면 온도를 조금 높게 설정한다.
3) 충전압력을 높게 한다.
4) 보압을 높인다.
5) 보압 유지시간을 더 길게 설정한다.
6) 충전량을 증가 시킨다.
7) 쿠션량을 증가 시킨다.
8) 사출성형기를 교체한다.
9) 사출장치의 역류 방지 링을 교체한다.
10) 사출성형기의 노즐 팁 직경을 더 크게 변경한다.

3. Countermeasures related to the design of the molded product

1) Increase the wall thickness of the molded product.
2) Provide ribs near the parts in which flow is difficult.

3. 성형 제품의 디자인 관련대책

1) 성형품의 두께를 늘린다.
2) 유동이 어려운 부분에 리브를 설치한다.

In the case of short shots due to air traps

(1) Countermeasures related to molds

① Provide an effective air vent in the part where air traps are occurring.
② Change the gate position.
③ Try changing the runner balance
④ Try making a structure in which the part with poor flow can be heated.
⑤ Try making the part with poor flow have a divided nested structure.

(2) Countermeasures related to the injection molding conditions

① Try changing the injection speed and changing the flow pattern
② Try changing the screw speed and the pressure selection position.
③ Try making the injection speed slower
④ Try setting the cavity surface temperature higher.
⑤ Try setting the mold clamping force a little lower

(3) Countermeasures related to the design of the molded product

① Investigate making the wall thickness of the molded product non-uniform.
② Increase the wall thickness of the molded product.

에어 트랩으로 인한 미성형의 경우

(1) 금형 관련 대책

① 에어 트랩이 발생하는 부분에 효과적인 에어벤트를 설치한다.
② 게이트 위치를 변경한다.
③ 런너 밸런스를 수정한다.
④ 흐름이 열악한 부분을 가열 할 수 있는 구조로 한다.
⑤ 불량한 흐름을 가진 부분을 분할된 구조로 만든다.

(2) 사출성형 조건에 관한 대책

① 사출 속도 변경 및 유동 패턴 변경한다.
② 스크류 속도와 압력 선택 위치를 변경한다.
③ 사출속도를 느리게 한다.
④ 캐비티 표면 온도를 더 높게 설정한다.
⑤ 형체력을 약간 낮춘다.

(3) 성형품의 설계에 관한 대책

① 성형품 살 두께의 불균일 여부를 조사한다.
② 성형품의 살 두께를 증가시킨다.

D 003 Solutions for Molding Defects (Flow Marks)

Flow marks are a phenomenon in which a pattern of the flow tracks of the molten plastic remains on the surface of the molded product. Depending on the extent, these can become defects in the case of molded products in which their external appearance is an important aspect of quality, such as in the case of home electrical appliances, containers for cosmetics, etc.

Flow marks are caused because there is a difference in the extent to which the cooling takes place upon contact with the surface of the mold at the front end of the plastic when the molten plastic is flowing inside the cavity of the mold.

The following countermeasures can be considered in order to avoid flow marks.

(1) Countermeasures related to molds

① Set the cavity surface temperature a little higher.
② Widen the gate.
③ Widen the runner.
④ Acquire sufficient cold slag.

(2) Countermeasures related to the injection molding conditions

① Increase the injection pressure.
② Increase the injection speed.
③ Acquire sufficient volume and increase the cushion amount.
④ Increase the dwell time.
⑤ Increase the plastic temperature.

D-003 성형 결함에 대한 해결책 (플로우 마크)

플로우 마크는 용융 수지의 유동 패턴이 성형품의 표면상에 남아있는 현상이다. 정도에 따라서는 가전제품, 화장품 용기 등 외관 품질이 중요한 성형품의 경우 결함이 될 수 있다.

용융수지가 금형의 캐비티로 유동 할 때 플라스틱의 유동선단이 금형의 표면과 접촉하면서 발생하는 냉각의 차이로 유동 흔적이 발생한다.

플로우 마크를 피하기 위해 다음과 같은 대책을 고려할 수 있다.

(1) 금형 관련 대책

① 캐비티 표면 온도를 조금 높게 설정한다.
② 게이트를 크게 한다.
③ 런너를 크게 한다.
④ 충분한 콜드 슬러그 웰을 확보한다.

(2) 사출성형 조건에 관한 대책

① 사출압력을 높인다.
② 사출속도를 높이다.
③ 충분한 사출량을 확보하고 쿠션량을 늘린다.
④ 보압 유지시간을 증가시킨다.
⑤ 용융 수지의 성형온도를 높인다.

(3) Countermeasures related to the design of the molded product

① Make variations in the wall thickness of the molded product small.

(3) 성형품의 설계에 관한 대책

① 성형품 살 두께 편차를 작게 한다.

D 004 Solutions for Molding Defects (Bubbles)

Bubbles (or voids) are a phenomenon in which air bubbles are left inside the molded product. In the case of transparent molded products such as lenses or prisms, bubbles become external appearance defects or defects in the optical characteristics. In mechanical parts, these lead to a reduction in strength or ultimately breakage.

The causes for the generation of bubbles can be broadly classified into two types.

One type is the bubbles that are caused by bubbles that got mixed with the molten plastic. These are called bubbles.

The other type is a vacuum void generated when the molded product shrinks. When sufficient dwell pressure did not act on the parts where the wall thickness of the molded product is thick, this phenomenon occurs simultaneously with the generation of sink marks due to abnormal shrinkage.

The following countermeasures can be taken for these types of defects.

D-004 성형 결함에 대한 해결책 (기포)

기포는 성형품 내부에 기포가 남는 현상이다. 렌즈나 프리즘과 같은 투명 성형품의 경우, 기포가 외관상의 결함이나 광학 특성의 결함이 된다. 기계 부품에는 강도가 저하되거나 궁극적으로 파손이 발생한다.

기포 발생 원인은 크게 두 가지 유형으로 분류 할 수 있다.

첫 번째 유형은 용융 된 플라스틱과 섞인 기포에 의해 발생한다. 이것을 버블이라고 한다.

또 다른 유형은 성형 제품이 수축 할 때 생성되는 진공 기포이다. 성형품의 살 두께가 두꺼운 부분에 충분한 보압이 작용하지 않을 때 발생하고, 이 현상은 비정상 수축에 기인하는 싱크 마크와 동시에 발생한다.

이러한 유형의 불량에 대해서는 다음과 같은 대응책을 취할 수 있다.

Countermeasures for Bubbles:

(1) Countermeasures related to molds

① There are no air vents, or they are insufficient.
② There is no cold slag well, or it is too small.

(2) Countermeasures related to the injection molding conditions

① The screw rotation speed is too high.
② The cylinder temperature is too high.
③ The injection speed is too high.

(3) Countermeasures related to the design of the molded product

① Insufficient preliminary drying of the molded material.

버블에 대한 대책 :

(1) 금형 관련 대책

① 에어벤트를 충분히 설치한다.
② 콜드 슬러그 웰을 크게 한다.

(2) 사출성형 조건에 관한 대책

① 스크류 회전속도를 낮춘다.
② 실린더 온도를 낮춘다.
③ 사출속도를 낮춘다.

(3) 성형 제품의 디자인 관련 대책

① 성형 재료의 충분한 예비 건조가 필요하다.

Countermeasures for Voids:

(1) Countermeasures related to molds

① There are no air vents, or they are insufficient.
② There is no cold slag well, or it is too small.
③ The sprue and the runners are too thin.
④ The gate is too small.

(2) Countermeasures related to the injection molding conditions

① The cavity surface temperature is too high.
② The dwell pressure is too low.
③ The dwell time is insufficient.

보이드에 대한 대책 :

(1) 금형 관련 대책

① 에어벤트를 충분히 설치한다.
② 콜드 슬러크 웰을 크게 한다.
③ 스프루와 런너를 크게 한다.
④ 게이트를 크게 한다.

(2) 사출성형 조건에 관한 대책

① 캐비티 표면 온도가 너무 높다.
② 보압이 너무 낮다.
③ 보압 유지시간이 충분하지 않다.

(3) Countermeasures related to the design of the molded product

① Insufficient preliminary drying of the molded material.
② The wall thickness of the molded product is too high.

(3) 성형 제품의 디자인 관련 대책

① 성형 재료의 충분한 예비 건조가 필요하다.
② 성형품의 살 두께를 줄인다.

D 005 Solutions for Molding Defects (Burn)

In plastic injection molding, some times burns occur at the end of thin ribs, etc., thereby causing a part of the molded product to become discolored black due to soot caused by burn.

The mechanism of burn is that, as the air inside the cavity of the mold is being vented out by the molten plastic that has entered the cavity it becomes trapped because there is no escape route for it, and because the air is compressed it generates heat and hence the plastic gets burnt due to the resulting heat that is generated. Since air is gaseous and generates heat when compressed, the trapped air generates heat. This is the same phenomenon as an air pump becoming hot when it is used for pumping air into a bicycle tire.

The compression of residual air inside the cavity is made in a very short time which is normally about 0.1 to 0.5 seconds, and also since it gets compressed to a very high pressure on the order of 200 to 500 kgf per square centimeter, the temperature rises easily to above the burning temperature of the plastic. (See Figure.)

D-005 성형 결함에 대한 해결책 (타버림)

플라스틱 사출성형에는 가끔 얇은 리브 등의 끝 부분에서 타는 현상이 발생하고, 탈 때 발생하는 그을음이 생겨서 성형품의 일부가 검게 변하게 된다.

타는 현상이 발생하는 것은 금형의 캐비티 내부의 공기가 캐비티에 들어간 용융수지에 밀려서 배출되지 못하고 갇히게 되어 압축되고, 열이 발생하기 때문에 플라스틱이 연소된다. 공기는 기체이므로 압축 될 때 열이 발생한다. 따라서 갇힌 공기에 열이 발생한다. 이것은 공기 펌프가 자전거 타이어로 공기를 펌핑 할 때 뜨거워지는 것과 같은 현상이다.

캐비티 내부의 잔류 공기의 압축은 일반적으로 0.1 ~ 0.5초의 매우 짧은 시간에 발생되는데, 그것은 평방 센티미터 당 200-500 kgf/cm²의 매우 높은 압력으로 압축되고, 플라스틱의 연소 온도 이상으로 온도가 쉽게 상승한다. (그림 참조).

Fig. Models of the flow of molten plastic into a rib

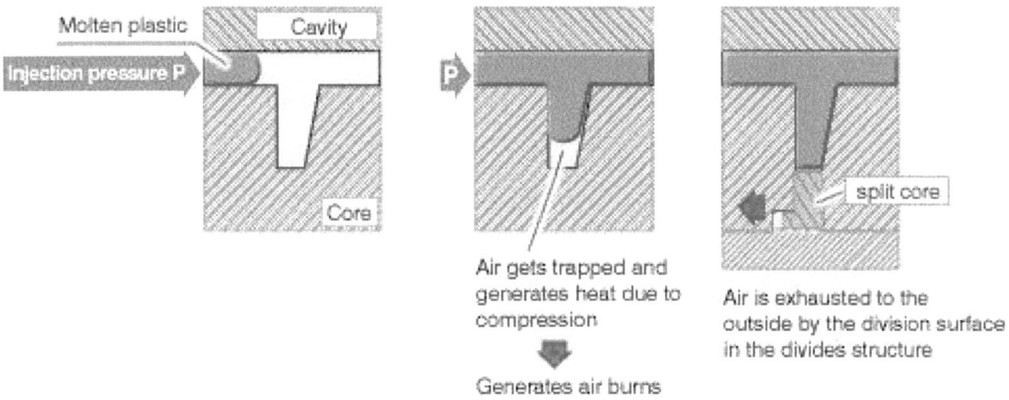

The following countermeasures are useful for preventing burn.

다음과 같은 대책은 타는 현상의 예방에 유용하다.

1) When the part into which plastic flows is closed, place a core pin as a split structure of the cavity. There will be no generation of burn since the air escapes to the outside through the gap between the cavity and the core pin. Providing an air vent on the side surface of the core pin will be more effective. However, since parting lines will appear on the surface of the molded product in the case of this method, care should be taken because this method cannot be used in the case of molded products on whose surface such parting lines can not allowed.

1) 플라스틱이 유동하는 부분의 코어 핀을 분할구조로 배치해라. 캐비티와 코어 핀 사이의 틈을 통해 공기가 외부로 빠져 나오기 때문에 타는 현상이 발생하지 않는다. 코어 핀의 측면에 에어벤트를 제공하는 것이 더 효과적이다. 그러나 이 방법의 경우 성형품 표면에 파팅라인 나타나기 때문에 파팅라인이 허용되지 않는 성형품의 경우에는 이 방법을 사용할 수 없으므로 주의해야한다.

2) In the molding conditions, make the injection speed as slow as possible and fill the cavity gradually. Although this improves the situation in the case of very light air burns, care should be taken because this is not a fundamental solution to the problem.

2) 성형 조건에서 가능한 천천히 사출속도를 설정하자. 이것은 타는 현상이 매우 작게 발생하는 경우는 개선되지만 문제의 근본적인 해결책이 아니기 때문에 주의를 기울여야 한다.

3) Carry out sufficient pre-drying of the material to be molded, and make sure that the condition is such that air does not get mixed inside the molten plastic. This too is not a fundamental solution to the problem so care should be taken.

3) 성형 할 재료를 사전에 충분히 건조시키고, 용융 수지 내부에의 공기가 섞이지 않도록 하자. 이것 역시 문제의 근본적인 해결책이 아니므로 주의를 기울여야 한다.

4) Change the wall thickness of the molded product or change the gate position thereby changing the flow pattern of the molten plastic and changing the position where air can get trapped. Although this method is effective, since the shape of the molded product and the weld position change, it is necessary to obtain the acceptance of the designer of the molded product.

4) 성형품의 살 두께를 변경하거나 게이트 위치를 변경하여 용융수지의 흐름 패턴을 변화시키고, 공기가 갇히게 될 수 있는 위치를 변경하자. 이 방법이 효과적이지만 성형품의 형상 및 웰드라인 위치가 변하기 때문에 성형품의 설계자와의 협의가 필요하다.

5) Change the injection speed selection position of the screw thereby changing the position where the air gets trapped. There are cases in which there are improvements using this method when the burn is light.

5) 스크류의 사출속도를 변경하여 공기가 갇히는 위치를 변경하자. 타는 현상이 약하게 발생할 때 이 방법을 사용하여 개선 된 경우가 있다.

A straight core pin for venting air is a new product that has the function of aiding the escape of air in the cavity or gas generated from the plastic to the outside at the time when plastic is being filled inside the cavity of the mold.

In simple terms, it is easy to understand this as a core pin with air vents provided on its side surfaces.

The key aspects of this product are summarized below

공기를 배출하기위한 직선형 코어 핀은 수지가 금형의 캐비티 내부로 채워지는 동안 캐비티에서 공기가 빠져 나가는 것을 돕거나 수지에서 발생하는 가스가 빠져 나가는 것을 돕는 신제품이다.

간단히 말해 측면에 에어벤트가 있는 코어 핀으로 쉽게 이해하면 된다.

이 제품의 주요 특징은 다음과 같다.

1) Since it is possible to set the tolerance of the diameter of the tip of the pin from the diameter of the pin body up to a maximum of 0.04 ㎜, depending on the type of plastic and the position at which the pin is assembled, it is possible to select the clearance of the air vent.

2) Since it is possible to select the effective length of the above small diameter tip part, it is possible to appropriately select this considering the relationship between the efficiency of the exhausting gas and the generation of burrs.

3) Since a deep gas vent groove has been provided in the middle of the pin, it is possible to exhaust the gas to outside the mold mainly from here.

4) There is a lineup of core pins whose diameters are from a minimum of 0.5 ㎜ to a maximum of 5 ㎜ and these can be used as core pins with relatively small diameters. It has been known from practical experience that the efficiency of gas exhaustion in precision fine molds has a big effect on maintaining the quality of molded products. In particular, in order to extend the maintenance cycle of molds, a very important point to which attention has to be paid is the venting of gas from small diameter bosses or hole shapes. In addition, even in the case of molds that carry out continuous molding using hot runners or valve gates, these type of gas vent pins are used very frequently in order to obtain stable quality.

5) It is also possible to add optional machining such as a flange cut to the core pin.

1) 핀이 조립되는 위치와 플라스틱의 종류에 따라 핀의 끝 직경에 0.04 ㎜의 허용 공차를 설정할 수 있으므로, 이 공차를 에어벤트로 사용가능하다.

2) 가는 직경 선단 부의 유효 길이를 선택할 수 있기 때문에 가스 빼기의 효율과 버(burr)의 발생의 관계를 고려하여 적절히 선택할 수 있다.

3) 핀의 중앙에는 깊은 가스 빼기 홈이 형성되어 있으므로 여기에서 주로 가스를 외부로 배출 할 수 있다.

4) 코어 핀 직경은 최소 0.5 ㎜에서 최대 5 ㎜ 이며, 상대적으로 작은 직경의 코어 핀들이 준비되어 있다. 실용적인 경험에 의하면, 정밀 미세 금형의 가스 빼기 효율은 성형품의 품질 유지에 큰 영향을 미친다고 알려져 있다. 특히 금형 유지보수 주기를 연장하기 위해서 주의가 요구되는 매우 중요한 점은 작은 지름의 보스 또는 구멍 모양에서 가스를 배출하는 것이다. 또한, 핫 런너 또는 밸브 게이트를 사용하여 연속 성형을 수행하는 금형의 경우에도 이러한 유형의 가스 빼기 핀은 안정된 품질을 얻기 위해 매우 빈번하게 사용한다.

5) 코어 핀에 플랜지 컷으로 선택 가공을 추가 할 수도 있다.

The following are the locations where air burns are likely to occur.

- At the tips of small diameter bosses on the back surface of case molded products.
- Tip parts of thin ribs.
- The end of ribs and bosses at locations where the wall thickness of the molded part has become thinner than other parts.
- The parts that are the farthest from the gate and that are filled last.
- Boundary ribs when square holes are next to each other.
- Molded products requiring high speed filling
- Molded products with thin walls.

다음은 공기에 의한 타는 현상이 발생할 수 있는 위치이다.

- 케이스 성형품 뒷면의 작은 지름의 보스 끝
- 얇은 리브의 끝 부분
- 성형품의 살 두께가 다른 부분보다 얇은 위치의 리브 및 보스의 끝
- 게이트에서 가장 멀리 있고, 마지막으로 채워지는 부품
- 사각 구멍이 서로 옆에 있을 때 리브 주위
- 고속 충전이 필요한 성형 제품
- 얇은 살 두께를 가진 성형 제품

When situations such as these have been recognized at the time of investigating the mold design, it is a very good practice to investigate the use of air venting straight core pins or nested divided structures starting in the design stage.

The following methods can be thought of as the countermeasures for solving the problem of burns.

금형설계를 조사 할 때 이러한 상황이 인식되면 설계 단계부터 공기 빼기용 직선 핀 또는 중첩 분할 구조의 사용을 고려하는 것이 좋다.

다음과 같은 방법은 타는 문제를 해결하기 위한 대책이다.

(1) Countermeasures related to molds

① Providing air vents.
② Making air vents deep, providing sufficient air exhaust path.
③ Taking apart the mold and cleaning and maintaining the air vents.
④ Using a divided structure or nested structure for the mold and letting the air escape from the parting surface.
⑤ Using a vacuum suction apparatus.
⑥ Changing the gate position.

(1) 금형 관련 대책

① 에어 벤트 부여
② 공기 배출구를 깊게 만들고 충분한 공기 배출 경로를 부여
③ 금형을 분해하고 에어 밴트 부를 청소 및 관리한다.
④ 금형에 분할 된 구조 또는 중첩 된 구조를 사용하고, 공기가 이형면에서 빠져 나가게 할 것
⑤ 진공 흡입 장치 사용
⑥ 게이트 위치 변경

(2) Countermeasures related to the injection molding conditions

① Making slower the injection speed.
② Making the cylinder temperature lower.
③ Making the charge quantity smaller.
④ Preventing stagnation of the plastic inside the cylinder.

(2) 사출성형 조건에 관한 대책

① 사출속도를 낮출 것
② 실린더 온도를 낮출 것
③ 충전량을 줄일 것
④ 실린더 내부 플라스틱의 정체를 방지할 것

(3) Countermeasures related to the molded product design

① Using a design that avoids thin walled parts in the product.
② Changing the wall thickness of the product, and changing the flow pattern of the molten plastic.
③ Changing the shape in which air gets trapped.

(3) 성형품 설계에 관한 대책

① 제품의 얇은 살 두께 설계를 피할 것
② 제품의 살 두께를 변경하고, 용융수지의 유동 패턴을 변경할 것
③ 에어 트랩이 발생할 형상을 변경할 것

D 006 Solutions for Molding Defects (Jetting)

Jetting (jet flow) is an external view defect in which a wavy pattern appears on the surface of the molded product. Jetting occurs because the plastic injected into the cavity from the gate suddenly flows into the cavity at a very high speed, and after colliding with the wall on the side opposite to the gate, proceeds to fill the cavity from the area surrounding the gate. Although the physical phenomenon is different from the injection molding formation of thermoplastic material, it is similar to the unstable and wavy ejection of toothpaste into the air when a tube of toothpaste is suddenly squeezed strongly. This can be thought of as the behavior of a viscous fluid.

D-006 성형 결함에 대한 해결책 (제팅)

제팅은 성형품의 표면에 물결 무늬가 나타나는 외관 결함이다. 제팅이 발생하는 이유는 게이트에서 캐비티로 주입 된 용융 수지가 갑자기 매우 빠른 속도로 캐비티로 흘러 들어가고 게이트 반대쪽 벽과 충돌 한 후 게이트 쪽으로 충전되기 때문이다. 비록 물리적인 현상은 사출성형과는 다르지만 치약 튜브가 갑자기 강하게 짜질 때 공기 중에 치약이 불안정하고 물결 모양으로 분출되는 것과 유사하다. 이것은 점성유체의 운동으로 생각할 수 있다.

The following methods can be considered as the countermeasures against jetting.

제팅에 대한 대책으로는 다음과 같은 방법이 있다.

1. Mold related countermeasures

1) Widen the gate. Make sure that the gate is not too thin compared to the wall thickness of the molded product.
2) Change the position of the gate, and move it to a location where successive filling can be done from near the gate.
3) Provide a core in the vicinity of the gate so as to obstruct the molten plastic that tries to suddenly fill the cavity.

1. 금형관련 대책

1) 게이트를 넓힌다. 게이트가 성형 제품의 살 두께에 비해 너무 얇지 않은지 확인해야 한다.
2) 게이트의 위치를 변경하고, 게이트 근처에서 점진적으로 충전이 가능한 위치로 이동한다.
3) 게이트 부근에 코어를 제공하여 갑자기 캐비티를 채우려는 용융 수지를 막을 수 있다.

2. Molding condition related countermeasures

1) Make the injection speed lower so that the filling is successively done from the vicinity of the gate.
2) As a corrective measure, set the cavity surface temperature a little higher so that the wavy patterns disappear. However, with this method, understand that the condition of the flow control is still unstable

2. 성형 조건 관련 대책

1) 사출속도를 낮추어 게이트 부근에서 점진적으로 충전되도록 하라.
2) 대책으로서 물결 모양 패턴이 사라지도록 캐비티 표면 온도를 조금 높게 설정한다. 그러나 이 방법을 사용하면 흐름 제어 상태가 여전히 불안정하다는 것을 이해해야한다.

D 007 Causes of and Countermeasures for Gas Burns

In plastic injection molding, a phenomenon can occur in which gas burns occur at the tip parts of thin ribs, etc., and a part of the molded product becomes black because of being carbonized.

The mechanism of generation of gas burns is that, at the time that the air inside the cavity of the mold is exhausted by the molten plastic flowing into the mold, if some of the air gets locked in a closed space, that air is compressed and gets heated by its own heat generation, thereby causing burning of the plastic.

Although air being a gas gets compressed, it generates heat when compressed. This is the same phenomenon as the pump becoming hot when it is used for filling air into a bicycle tire.

The compression of the air remaining inside the cavity occurs in a very short time of about 0.1 to 0.5 second, and since it is also compressed to a very high pressure such as 200 to 500 kgf/cm^2 , the temperature easily reaches the burning temperature of the plastic material. (See Fig.)

D-007 가스에 의해 타는 원인 및 대책

플라스틱 성형에서는 얇은 리브 등의 끝단부에 가스에 의한 타는 문제가 발생하고, 성형품의 일부가 탄화되어 흑색으로 변하는 현상이 발생할 수 있다.

가스에 의해 타는 현상의 메커니즘은 용융수지가 금형에 충전 될 때, 캐비티 내부의 일부 공기가 폐쇄 된 공간에 갇히게 되면 그 공기가 압축되고, 자체 발열에 의해 가열되어 성형품에 연소가 발생한다.

기체인 공기는 압축되면 열이 발생한다. 이것은 자전거 타이어에 공기를 채우는데 사용되는 펌프가 뜨거워지는 것과 동일한 현상이다.

캐비티 내부에 남아있는 공기의 압축은 약 0.1 ~ 0.5초의 매우 짧은 시간에 발생하고, 200 ~ 500 kgf/cm^{2}와 같은 매우 높은 압력으로 압축되기 때문에 플라스틱을 쉽게 타게 하는 온도까지 상승한다.(그림 참조)

[Fig.2] Seal Tape Termination

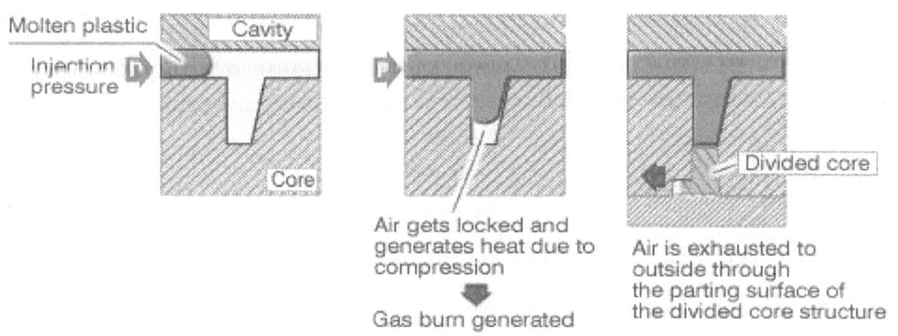

The following countermeasures are effective for preventing gas burns.

다음 대책은 가스에 의해 타는 현상을 방지하는데 효과적 이다.

1) If the part into which plastic flows is in the closed state, use a divided structure and insert a core pin. Since air is exhausted between the cavity and the core pin, gas burns are not generated. It is more effective if an air vent is provided on the side surface of the core pin.
However, since in this method, parting line will appear on the surface of the molded item, it is necessary to take care because this method cannot be used in the case of molted items that do not permit such parting lines on the surface.

1) 용융수지가 유동하는 부분이 닫힌 상태 인 경우 분할구조와 코어 핀을 삽입하라. 캐비티와 코어 핀 사이에서 공기가 배출되므로 가스에 의한 타는 문제가 발생하지 않는다. 코어 핀의 측면에 에어벤트가 설치되어있는 쪽이 효과적이다.
그러나 이 방법은 성형품 표면에 파팅 라인이 생기기 때문에 표면에 파팅 라인을 허용하지 않는 성형품의 경우에는 이 방법을 사용할 수 없기 때문에 조심해야한다.

2) In the molding conditions, make the injection speed as low as possible and fill slowly. Although this may improve the situation in the case of light gas burns, it is necessary to be aware that this is not a fundamental countermeasure.

2) 성형 조건에서 사출속도를 가능한 낮추고 천천히 한다. 이것은 약하게 가스 타는 경우를 개선할 수 있지만, 근본적인 대책은 아니다.

3) Carry out the preliminary drying of the molding material sufficiently, and put it in a condition in which it is difficult for air to get mixed inside the molten plastic.
Even in this case, it is necessary to be aware that this is not a fundamental countermeasure.

3) 성형 재료의 예비 건조를 충분히 실시하고, 용융수지 내부에서 공기가 혼합되지 않게 해야 한다.
이 경우도 근본적인 대책이 아니라는 사실을 알아야 한다.

4) Change the flow pattern of the molten plastic by changing the wall thickness of the molded product, or by changing the gate position, thereby changing the position where air gets trapped. Although this method is effective, since the shape of the molded product and the weld position change, it is necessary to obtain permission in advance from the designer of the molded product.

4) 성형품의 살 두께를 변경하거나 게이트 위치를 변경하므로 용융수지의 흐름 패턴을 변경하여 공기가 갇히는 위치를 변경하여야 한다. 이 방법이 효과적이지만, 성형품의 형상 및 웰드 라인의 위가가 변화하기 때문에 성형품의 설계자로부터 미리 허락을 받아야한다.

5) Change the position where the air gets trapped by changing the position of changing the injection speed of the screw.

This may improve the situation in some cases if the gas burns are light.

5) 사출성형기 스크류의 사출속도를 변경하여 공기가 갇히는 위치를 변경하자.

이 경우는 가스에 의하여 타는 현상이 크지 않는 상황에서 개선할 수 있다.

D 008 Solutions for Molding Defects (Silver Streaks)

D-008 성형 결함에 대한 해결책(실버 스트리크)

Silver streaks are a phenomenon in which shining line shaped patterns appear on the surface of the molded product. These can be considered external appearance quality defects in exterior parts of consumer electrical products, automobiles, motorcycles, etc.

The cause of silver streaks is the air or gases contained in the molding material appearing on the surface of the molded product.

The following methods can be thought of as the countermeasures for solving the problem of silver streaks.

실버 스트리크는 성형품 표면에 은색의 빛나는 줄무늬가 나타나는 현상이다. 이는 전기 제품, 자동차, 오토바이 등의 외관 부품의 결함으로 간주 될 수 있다.

성형품 표면에 나타나는 실버 스트리크의 원인은 성형재료에 포함된 공기 또는 가스 때문이다.

실버 스트리크의 문제를 해결하기 위한 대책으로 다음과 같은 방법을 생각할 수 있다.

1. Countermeasures related to molds

1) Enhancing the function of air vents.
2) Making gate size larger.
3) Making the cold slag well larger

1. 금형 관련 대책

1) 에어 벤트 기능의 강화
2) 게이트 크기를 크게 함
3) 콜드 슬러그 웰을 크게 함

2. Countermeasures related to the injection molding conditions

1) Checking the preliminary drying conditions (temperature, drying time) of the molding material, and carrying out appropriate drying.

This is the most important countermeasure.

① Making the injection speed lower and filling the cavity slowly.
② Making the cylinder temperature lower.
③ Making the screw rotation speed lower.
④ Preventing stagnation of the plastic inside the cylinder.

2. 사출성형 조건에 관한 대책

1) 성형 재료의 예비 건조 조건 (온도, 건조시간) 확인 및 적절한 건조 수행

이것이 가장 중요한 대책이다.

① 사출속도를 낮추고, 천천히 충전 할 것
② 실린더 온도를 낮출 것
③ 스크류 회전속도를 낮출 것
④ 실린더 내부의 용융수지의 정체 현상을 방지할 것

3. Countermeasures related to the molded product design

1) Making the wall thickness of the molded product as uniform as possible.

3. 성형품 설계에 관한 대책

1) 성형품 살 두께의 균일화

D 009 Solutions for Molding Defects (Gloss Defect)

Although it is ideal that the surface of the molded product is a copy of the surface of the cavity and has a beautiful shape and gloss, some times the surface gloss may become clouded, or there may be variations in the gloss.

D-009 성형 결함에 대한 해결책 - 광택 결함

성형품의 표면은 캐비티 표면의 복사물이고, 아름다운 모양과 광택을 가지는 것이 이상적이지만 때때로 표면 광택이 흐려지거나 광택에 변화가 있을 수 있다.

Particularly in the case of products with decorative surfaces in which the external appearance quality is very important, gloss defect can become a very serious defect.

The following methods can be thought of as the countermeasures for solving the problem of gloss defects.

특히, 외관 품질이 매우 중요한 장식용 제품의 경우, 광택 결함은 매우 심각한 결함이 될 수 있다.

다음의 방법들은 광택 결함의 문제를 해결하기 위한 대책으로 고려할 수 있다.

1. Countermeasures related to molds

1) There are no air vents or they are insufficient.
2) The gate is small.
3) The sprue and runners are thin.
4) The electroplating condition of the cavity surface is not good.
5) The polishing condition of the cavity surface is not good.
6) There are some deposits (soot) adhered to the cavity surface.

1. 금형 관련 대책

1) 충분한 에어벤트를 설치할 것
2) 게이트를 크게 할 것
3) 스프루와 런너의 직경을 크게 할 것
4) 캐비티 표면의 전기 도금 조건의 개선
5) 캐비티 표면의 연마 조건의 개선
6) 캐비티 표면에 부착 된 찌꺼기의 제거

2. Countermeasures related to the injection molding conditions

1) The charge quantity is insufficient.
2) The cushion position is too short
3) The holding pressure is too low.
4) The pressure holding time is too short.
5) The cavity surface temperature is low.
6) The preliminary drying of the molding material is insufficient.

2. 사출성형 조건에 관한 대책

1) 충분한 충전량을 제공할 것
2) 쿠션 위치를 길게 할 것
3) 보압력을 크게 할 것
4) 보압 유지시간을 길게 할 것
5) 캐비티 표면 온도를 높일 것
6) 충분한 성형 재료의 예비 건조

D 010 Mold Galling

As the mold for plastic injection molding is continued to be used, the slide core or ejector pin, ejector sleeve, and center pin, etc., can cause what is called "galling". "Galling" is the abnormal wear of sliding surfaces, and the causes of abnormal wear are classified as follows.

(1) Abrasive wear

This is a form of abnormal wear that can occur easily if there is a difference in the hardness of the materials of the sliding mold parts. This is the phenomenon of a harder material scratching the softer material and getting fused to it.

(2) Adhesion wear

In this condition, the projecting parts of mold parts hit against each other and become adhered to each other at the locations where the contact is very severe, and the adhered part falls off becoming wear dust, and the wear continues.

(3) Fatigue wear

Fatigue occurs when a mold part repeatedly moves and stops, and this causes wear in this condition. This is the condition when flaking has taken place (peeling off of scales).

D-010 금형의 긁힘

슬라이드 코어 또는 이젝터 핀, 이젝터 슬리브 및 센터 핀 등이 지속적으로 작동하는 금형에 "긁힘"이 발생할 수 있다. "긁힘"은 슬라이딩 동작이 발생하는 면의 비정상적인 마모이고, 이러한 비정상적 마모의 원인은 다음과 같이 분류된다.

(1) 연마 마모

이것은 금형의 슬라이딩 작동을 하는 부품의 경도 차이가 있을 때 쉽게 발생하는 비정상적 마모의 한 형태이다. 이 현상은 경도가 높은 재료가 경도가 낮은 재료를 긁어내고, 점차 마모시키는 것이다.

(2) 접착 마모

금형 부품의 돌출된 부분이 다른 부품과 접촉하고, 접촉이 매우 심한 부분에는 서로 접착되고, 접착 된 부품이 떨어져 마모 분진이 되어 마모가 계속된다.

(3) 피로 마모

피로는 금형 부품이 빠르게 움직이고 정지할 때 발생하며, 이런 움직임이 마모의 원인이 된다. 얇게 갈라져 떨어진다 (비늘 벗기기).

(4) Fretting Corrosion

This is a form of wear in which wear of the pitching shape occurs in the regions where parts mate with each other with a relatively small clearance. This occurs very often in square keys and key grooves.

(4) 침식 부식

이것은 비교적 작은 간극을 두고 서로 짝을 이루는 부품 영역에서 피칭(반복돌기무늬) 형태의 마모가 발생하는 형태이다. 이것은 사각 키 및 키 홈에서 매우 자주 발생한다.

(5) Corrosion wear

In a corrosive atmosphere of chemical constituents, moisture, or ions, etc., arising from the plastic, wear of the mold parts occurs due to the generation of a potential difference between the two.

When galling occurs, fatal damage is caused to the cavity and the core, and when there is defective movement of the ejector pin or the slide core, it is highly likely to result in breakage of the mold. In order to prevent galling, it is necessary to carry out the appropriate lubrication control, and to use mold parts that are maintenance free.

(5) 부식 마모

플라스틱으로 인해 발생하는 화학 성분, 수분 또는 이온 등의 부식성 분위기에서 두 부분 사이에 전위차가 생겨서 금형 부품이 마모된다.

긁힘이 발생하면 캐비티와 코어에 치명적인 손상이 발생하고, 이젝터 핀이나 슬라이드 코어의 움직임이 불량하면 금형이 파손될 가능성이 높다. 긁힘을 방지하기 위해서 적절한 윤활 작업을 수행하고, 유지보수가 필요 없는 금형 부품을 사용해야한다.

D 011 Wear of Molds

The parts of plastic injection mold wear out due to contact or friction between parts, and in addition, due to glass fibers contained in the plastic resin.

If the shape of the wear exceeds the tolerable range, problems occur such as the mold can no longer move correctly, the mold will be broken, or the shape of the molded product becomes deformed.

D-011 금형의 마모

플라스틱 사출금형 부품은 플라스틱 수지에 포함 된 유리 섬유와의 접촉과 마찰로 인해 마모된다.

마모의 형상이 허용 범위를 초과하는 경우 금형이 더 이상 정확하게 동작하지 않거나, 금형의 파손 또는 성형품의 형상이 변형되는 등의 문제가 발생한다.

The worn out of molds is classified into normal wear and abnormal wear.

Normal wear is the worn out that is caused when parts that touch or slide against each other gradually get worn out.

Although it is technically possible to make it difficult for parts to get worn out, it is extremely difficult to prevent wear fundamentally unless the parts do not contact each other.

Normal wear is classified into initial wear and steady wear.

If the part is replaced with a new part when the steady wear reaches managed scheduled dimensions, it is possible to prevent in advance failures or problems with the molds.

On the other hand, abnormal wear is wear that is not normal wear. There are five typical classifications of abnormal wear.

금형의 마모는 정상 마모와 비정상 마모로 나눈다.

일반적인 마모는 서로 닿거나 미끄러지는 부분이 점차적으로 마모되면서 발생한다.

기술적으로 부품에 마모가 발생하지 않도록 하는 것은 가능하지만, 부품이 서로 접촉하지 않는다면 근본적으로 마모를 방지하는 것은 극히 어렵다.

보통 마모는 초기 마모와 지속적 마모로 분류된다.

지속적 마모가 발생하는 부품은 예정 관리 치수에 도달 할 때 새 부품으로 교체하면 사전에 금형 불량 또는 문제를 예방할 수 있다.

반면 비정상적인 마모는 일반적인 마모가 아니다. 비정상적인 마모에는 5 가지 전형적인 분류가 있다.

(1) Abrasive wear

This is the form of wear that can occur easily when there is a difference in the hardness of the materials that are rubbing against each other. The harder material bites into the softer material generating scratches, and causing wear.

(1) 연마 마모

이것은 서로 마찰하는 재료의 경도에 차이가 있을 때 쉽게 발생할 수 있는 마모이다. 더 견고한 소재는 부드러운 소재에 맞닿아 긁힘을 유발시키고, 이는 마모의 원인이 된다.

(2) Adhesive wear

This is the form of wear in which projecting parts of materials hit against each other causing a part to get adhered, and as a result of growth of the adhered part, it becomes a transfer particle, and eventually falls off as wear dust.

(2) 접착 마모

이것은 재료의 돌출 부분이 서로 충돌하여 부품이 달라붙는 현상이며, 접착 부분의 성장으로 인하여 입자가 되어 결국 마모 분진이 된다.

(3) Fatigue wear

This is the form of wear in which metal fatigue occurs due to repletion of the application and removal of load (repetition of operation and stopping), causing wear.

(3) 피로 마모

이것은 (작동 및 정지의 반복) 하중의 가해와 제거가 반복됨으로 인한 금속의 피로로 발생하는 마모이다.

(4) Fretting wear

This is the form in which wear of fine pitching shape occurs on the surface of mating parts.

(4) 프레팅(침식) 마모

이것은 상대 부품의 표면에 미세 피칭 형상(반복 돌기무늬)의 마모가 발생하는 형태이다.

(5) Corrosion wear

This is the form of wear that is caused when a potential difference is generated between metals in a corrosive atmosphere, thereby causing the sliding parts to disappear, resulting in damage occurring speedily due to the addition of friction wear.

(5) 부식 마모

부식성 분위기의 금속 사이에서 발생하는 부식으로 슬라이딩 부분이 없어지므로서 마찰 마모로 인한 추가적인 손상이 빠르게 발생하게 된다.

E. 플라스틱수지 특성

CHAPTER E

E 001 Molding Shrinkage Ratios of Major Plastic Materials

E-001 주요 플라스틱 재료의 수축률

In order to carry out the design of molds for plastic injection molding, it is necessary to determine the molding shrinkage ratio. In this course, rough guides of the molding shrinkage ratios are explained for the typical plastic materials used in injection molding.

[Table 1] is a list of the major thermoplastic materials, their molding shrinkage ratios, cavity surface temperatures, and injection molding pressures.

For more details, it is common practice to obtain the material catalogs and technical documents for each grade from the manufacturers of the molding materials and to use them as the reference materials for making decisions.

The values given here are for natural materials unless specified otherwise.

플라스틱 사출금형의 설계를 진행하기 위해서 성형 수축률을 결정할 필요가 있다. 이 과정에서는 사출성형에 사용되는 일반적인 플라스틱 재료에 대한 대략적인 성형 수축률 가이드를 설명한다.

[표 1]은 주요 열가소성 재료, 성형 수축률, 캐비티 표면 온도 및 사출성형 압력을 나타낸 것이다.

자세한 내용은 성형 재료 제조업체로부터 각 그레이드에 대한 재료 카탈로그 및 기술 문서를 얻고, 결정을 내릴 때 참고 자료로 사용하는 것이 일반적이다.

여기에 주어진 값은 달리 명시되지 않는 한 첨가제가 없는 천연 수지에 대한 값이다.

[Table 1] List of molding shrinkage ratios of major plastic materials

Plastic material name	Shrinkage ratio (%)	Cavity surface temperature (℃)	Injection molding pressure		Plastic material name	Shrinkage ratio (%)	Cavity surface temperature (℃)	Injection molding pressure	
			(kgf/cm²)	(MPa)				(kgf/cm²)	(MPa)
ABS Acrylonitrile Butadiene Styrene polymer	0.4~0.9	50~80	550~1750	53.97~171.7	POM Poly acetal	2.0~2.5	60~120	700~1400	68.69~137.4
PS Polystyrene	0.4~0.7	20~60	700~2100	68.69~206.1	Poly butylenes terephthalate PBT with 30% glass fibers	0.2~0.8	40~80	560~1800	54.95~176.6
AS Acrylonitrile styrene	0.2~0.7	50~80	700~2300	68.69~225.7	PC Polycarbonate	0.5~0.7	80~120	700~1400	68.69~137.8
EVA Ethylene vinyl acetate	0.7~1.2	50~80	1050~2800	103~274.8	Poly phenylene sulfide PPS with 40% glass fibers	0.2~0.4	130~150	350~1400	34.34~137.8
PP Poly propylene	1.0~2.5	20~90	700~1400	68.69~137.8	Liquid crystal polymer LCP with 40% glass fibers	0.2~0.8	70~110	700~1400	68.69~137.8
Poly propylene with 40% glass fibers	0.2~0.8	20~90	700~1400	68.69~137.8	Modified polyphenylene oxide (Modified PPO)	0.1~0.5	80~90	-	-
DPE High density polyethylene	2.0~6.0	10~60	700~1400	68.69~137.8	PSF Poly sulfone	0.7~0.8	90~100	-	-
PmmA Methacrylic acid methyl ester (acrylic)	0.1~0.4	40~90	700~1400	68.69~137.8	PES Polyether sulfone	0.6~0.8	120~140	-	-
PA6 Polyamide (Nylon 6)	0.5~1.5	40~120	350~1400	34.34~137.4	PET Poly ethylene terephthalate	0.2~0.4	70~100	-	-
PA66 Polyamide (Nylon 66)	0.8~1.5	30~90	350~1400	34.34~137.4	PEEK Polyether ether ketone	0.7~1.9	120~160	-	-

E 002 Guide to the Molding Shrinkage ratio of Major Plastic resins (1)

It was already explained that it is very important to determine the molding shrinkage ratio for designing the molds for plastic injection molding. Here we would like to explain some guides to the molding shrinkage ratio of some typical plastic resins.

Table 1 shows some major thermoplastic plastic resins and their molding shrinkage ratios, cavity surface temperatures, and injection molding pressures. For more details, it is common to obtain the material catalogs or technical documents for each grade of material from the manufacturer of the plastic resin and use those documents for making this decision.

* Unless otherwise stated, the values given here are for natural resins.

E-002 주요 플라스틱 수지의 성형 수축률에 대한 가이드(1)

사출금형 설계 시 성형 수축률을 결정하는 것이 매우 중요하다고 이미 설명했다. 여기에 몇 가지 일반적인 플라스틱 수지의 성형 수축률에 대한 가이드를 설명하고자 한다.

표 1은 주요 열가소성 플라스틱 수지 및 성형 수축률, 캐비티 표면온도 및 사출성형 압력을 나타낸다. 자세한 내용은 플라스틱 수지 제조업체로부터 각 그레이드에 대한 재료 카탈로그 또는 기술 문서를 통하여 결정하는 것이 일반적이다.

* 특별한 언급이 없는 한, 여기에 주어진 값은 일반 수지에 대한 값이다.

Plastic resin name	Molding shrinkage ratio (%)	Cavity surface temperature (℃)	Injection molding pressure	
			(kgf/sq.cm)	(MPa)
Acrylonitrile-styrene copolymer ABS	0.4 ~ 0.9	50 ~ 80	550 ~ 1750	53.97 ~ 171.7
Polystyrene PS	0.4 ~ 0.7	20 ~ 60	700 ~ 2100	68.69 ~ 206.1
Acrylonitrile-styrene AS AS	0.2 ~ 0.7	50 ~ 80	700 ~ 2300	68.69 ~ 225.7
Ethylene vinyl acetate EVA	0.7 ~ 1.2	50 ~ 80	1050 ~ 2800	103 ~ 274.8
Polypropylene PP	1.0 ~ 2.5	20 ~ 90	700 ~ 1400	68.69 ~ 137.8
Polypropylene glass fiber 40%	0.2 ~ 0.8	20 ~ 90	700 ~ 1400	68.69 ~ 137.8

High Density Polyethylene HDPE	2.0 ~ 6.0	10 ~ 60	700 ~ 1400	68.69 ~ 137.8
Methacrylic acid methyl ester (acrylic) PMMA	0.1 ~ 0.4	40 ~ 90	700 ~ 1400	68.69 ~ 137.8

E 003 Guide to the Molding Shrinkage ratio of Major Plastic Resin (2)

Although the guidelines for the molding shrinkage ratios of typical plastic resins were explained in the last lesson, we would like to add some more information. This time, we provide the values for some engineering plastics resins.

The table shows the major thermoplastic resins and their molding shrinkage ratios, cavity surface temperatures, and injection molding pressures. For more details, it is common to obtain the material catalogs or technical documents for each material grade from the manufacturer of the material, and to use them as the documents for making decisions.

E-003 주요 플라스틱 수지의 성형 수축률에 대한 가이드(2)

일반적인 플라스틱 수지의 성형 수축률에 대한 지침이 마지막 강의에서 설명되었지만 몇 가지 추가 정보를 더하고자 한다. 이번에는 일부 엔지니어링 플라스틱 수지에 대한 값을 제공한다.

표에는 주요 열가소성 수지 및 성형 수축률, 캐비티 표면 온도 및 사출성형 압력이 나와 있다. 자세한 내용은 재료 제조업체의 각 재료 등급에 대한 재료 카탈로그 또는 기술 문서를 참고하여 결정하는 것이 일반적이다.

Table of molding shrinkage ratios of major plastic resins (2)

Name of plastic resin	Molding shrinkage ratio (%)	Cavity surface temperature (℃)	Injection molding pressure	
			(kgf/cm²)	(MPa)
Polyamide (Nylon 6) PA6	0.5 - 1.5	40 - 120	350 - 1400	34.34 - 137.4
Polyamide (Nylon 66) PA66	0.8 - 1.5	30 - 90	350 - 1400	34.34 - 137.4
Polyacetal POM	2.0 - 2.5	60 - 120	700 - 1400	68.69 - 137.4
Polybutylene terephthalate PBT 30% glass fiber	0.2 - 0.8	40 - 80	560 - 1800	54.95 - 176.6
Polycarbonate PC	0.5 - 0.7	80 - 120	700 - 1400	68.69 - 137.8
Polyphenylene sulfide PPS 40% glass fiber	0.2 - 0.4	130 - 150	350 - 1400	34.34 - 137.8
Liquid crystal polymer LCP 40% glass fiber	0.2 - 0.8	70 - 110	700 - 1400	68.69 - 137.8

E 004 Appropriate Pre-drying of Plastic Molding Materials

Usually, the plastic materials are formed in the shape of pellets and sent from the raw material manufacturers in paper bags, etc.

Since these pellets would have absorbed the moisture in the atmosphere, if they are used for injection molding while they still contain a lot of moisture, depending on the type of material, they can undergo hydrolysis, or their physical properties can decrease. Further, it is possible for silver streaks to appear on the surface of

E-004 플라스틱 성형 재료의 적절한 사전 건조

일반적으로 플라스틱 재료는 펠릿의 형태로 종이 자루 등에 담겨 원료 제조업체에서 보내진다.

이 펠릿은 대기 중의 수분을 흡수하였을 것이므로 만약, 사출성형에 사용되는 수지가 수분을 많이 함유한 상태에서 사출성형에 사용하면 가수 분해되거나 물리적 특성이 저하 될 수 있다. 또한, 성형품의 표면에 실버 스트리크가 나타나기

the molded product, and it is also possible for short shots or burn due to gas to occur.

In view of this, in the case of most molding materials, before they are put into the hopper drier, it is necessary to carry out pre-drying in a box type drying oven.

In pre-drying it is recommended to observe the appropriate drying temperature and drying time. This is because, if the drying is done at a temperature less than the appropriate temperature, even if the drying is done for a long time, the moisture content cannot be removed as desired. A material whose pre-drying has been completed should be used as quickly as possible. When any left over material is to be used some days later, carry out its pre-drying again before use.

The pre-drying conditions of special plastic are listed in Table 1.

쉽고, 가스에 의한 미성형 또는 타는 문제가 발생할 수 있다.

이러한 점에서 대부분의 성형 재료의 경우 호퍼 드라이어에 넣기 전에 상자형 건조 오븐에서 예비 건조를 해야 한다.

사전 건조 시에 적절한 건조 온도와 건조시간을 준수하는 것이 좋다. 이는 적절한 온도보다 낮은 온도에서 건조를 행하면, 장시간 건조해도 원하는 수분을 제거 할 수 없기 때문이다. 사전 건조가 완료된 재료는 가능한 빨리 사용해야 한다. 며칠 동안 재료를 남겨두면 사용하기 전에 다시 건조해야 한다.

특수 플라스틱의 사전 건조 조건은 표 1에 나열되어있다.

Table 1 Pre-drying temperatures of plastic molding materials

Material name	Symbol	Pre-drying temperature (℃)	Drying time (H)	Material name	Symbol	Pre-drying temperature (℃)	Drying time (H)
Liquid crystal polymer	LCP	110~150	4~8	ABS	ABS	70~80	2~3
Polyether imide	PEI	120~150	2~7	Acrylic	PmmA	70~100	2~6
Polyamide imide	PAI	150~180	8~16	Polycarbonate	PC	120	4~6
Thermoplastic elastomer	TPE	120	3~4	Nylon 6	PA6	80	8~15
Poly ether ether ketone	PEEK	150	8	Nylon 66	PA66	80	8~15
Poly phenylene sulfide	PPS	140~250	3~6	Nylon 11	PA11	70~80	8~15
Poly allylate	PAR	120~150	4~8	Nylon 46	PA46	80	8~10
Poly sulfone	PSU	120~150	3~4	Poly acetal	POM	110	2~3
				PBT	PBT	120	4~5

Pellets of plastic molding materials generally absorb moisture from the atmosphere to a certain extent. If the quantity of absorbed moisture is large, the plastic can undergo hydrolysis in the process of being melted and mixed in the cylinder of the injection molding machine, or, when molding is done, this can cause silver streaks, air bubbles, or glossiness defects on the surface of the molded product, or can cause copying defects, etc. Therefore, it is necessary beforehand to put the pellets of molding materials in a drying apparatus and remove the moisture content in them. If the pre-drying is not done appropriately, it can lead to variations in the fluidity, deterioration of physical characteristics, and molding defects.

The following are the main types of driers being used at present.

플라스틱 성형 재료의 펠릿은 일반적으로 대기로부터 일정 수준의 수분을 흡수한다. 흡수 된 수분의 양이 많으면 플라스틱은 사출성형기의 실린더 내에서 용융 및 혼합되는 과정에서 가수 분해되거나, 성형품 표면에 실버스트리크, 기포 또는 광택 결함이 생기거나 전사 결함 등을 일으킬 수 있다. 따라서 성형 재료의 펠릿을 건조 장치에 넣고 그 내부의 수분 함량을 제거하는 것이 필요하다. 예비 건조가 적절하게 이루어지지 않으면, 유동성의 변화, 물리적 특성의 악화 및 성형 결함을 초래할 수 있다.

다음은 현재 사용 중인 건조기의 주요 유형이다.

1. Hot air drier

The hopper drier and the box type drier are the typical equipment used with this type. The drying method is that of blowing hot air at the pellets thereby evaporating the moisture in them. Although this is a co mmon and simple drying method, this method is not suitable for completely removing the moisture content.

1. 열풍 건조기

호퍼 드라이어 및 박스형 드라이어가 이 유형에 속하는 전형적인 장비이다. 건조 방법은 펠릿에 뜨거운 공기를 불어 넣어서 습기를 증발시키는 것이다. 이것은 일반적인 건조 방법이지만 수분을 완전히 제거하는 데는 적합하지 않다.

2. Dehumidified hot air drier

In this method, after first removing the moisture in air, that air is heated and blown on the pellets thereby evaporating the moisture content in them. Since the air used for drying is re-circulated and used again after being dehumidified, heat loss will be small, and it is possible to carry out rational drying. This method is suitable for drying PBT, etc.

2. 열풍 제습 건조기

이 방법에는 먼저 공기 중의 수분을 제거한 후 그 공기를 가열하여 펠릿에 불어 넣음으로써 내부의 수분을 증발시킨다. 건조에 사용 된 공기는 제습 후에 재순환되고 다시 사용되기 때문에 열 손실이 적어 합리적인 건조가 가능하다. 이 방법은 PBT 건조 등에 적합하다.

3. Reduced pressure heat transfer type drier

This is a method for evaporating the moisture content in the pellet by heat transfer in a reduced pressure environment. Drying at low temperatures becomes possible, and hence it is possible to prevent the oxidization of plastic and to reduce the effects of additives in the pellet. In addition, this is also a method in which thermal loss is also small. This type of drier is attracting a lot of attention as the drier of the future.

*Reference: "Injection Molding Dictionary", p. 214, Peripheral and accessory equipment (Hideki Kubo, Industry Research Institute, (2002))

3. 감압 열전달식 건조기

이것은 감압 환경에서의 열전달에 의해 펠릿 내의 수분 함량을 증발시키는 방법이다. 저온에서의 건조가 가능해지고, 플라스틱의 산화를 방지하고 펠릿 내의 첨가제의 효과를 감소시킬 수 있다. 또한 열손실이 적은 방법이기도 하다. 이 유형의 건조기는 미래의 건조기로서 많은 주목을 받고 있다.

* 참고 문헌 : "Injection Molding Dictionary", p. 214, Peripheral and accessory equipment (Hideki Kubo, Industry Research Institute, (2002))

E 005 Types of Plastics and Guidelines for Selecting Steel for the Cavity

When the steel material for the cavity is selected depending on the type of plastic resin and the functions expected of the molded product.

If the selection of the steel is not appropriate, it may not be possible to polish a fine surface, or corrosion may progress during the molding, or the wearing out may occur at a fast rate.

The table below shows the major types of plastic resins or the guidelines for selecting the recommendation of steel.

E-005 플라스틱의 종류와 캐비티용 강재의 선택을 위한 가이드 라인

캐비티용 강재는 플라스틱 수지의 종류 및 성형품에 요구되는 기능에 따라 선택한다.

강재의 선택이 적절하지 않은 경우 정밀한 표면으로 연마하는 것이 불가능하거나, 성형 중에 부식이 진행되거나, 또는 마모가 빠른 속도로 발생할 수 있다.

아래 표는 주요 플라스틱 수지와 강재를 선택하기 위한 지침을 나타낸다.

	Plastic resin	Example of molded product	Characte-ristics expected of the plastic resin	Characte-ristics expected of the steel	Recommend-ed type of steel
Thermoplastic Resin	PP ABS	Bumpers OA chassis	Shock resistance	Surface gloss	S50C SCM440
	PS PMMA ABS	Lighting fixtures General goods Containers for cosmetics	Design flexibility	Surface gloss Mirror finishing	SKD61 Pre-hardened steel
	POM PA	Gear wheels Bearings	Abrasion resistance	Abrasion resistance	SKD61 Pre-hardened steel
	PC PMMA	Lenses Light guides	Transparency Optical transmittivity	Mirror finishing	SUS420J2 Precipitation hardened steel
	PC PMMA	CDs DVDs	Optical transmittivity Optical birefringence characteristics	Mirror finishing Corrosion resistance	SUS420J2
	PVC	Rainwater pipes	Flaming resistance	Corrosion resistance	SUS
	Fire resistant ABS	TV cabinets Consumer electronics components	Flaming resistance	Corrosion resistance	SUS420J2 Pre-hardened steel
	PBT-GF PA-GF	Camera chassis Electrical components	Durability	Abrasion resistance	Pre-hardened steel SKD11
	PA with magnetic powder added	Printer rollers Sensor components	Formability Electrical characteristics	Non-magnetic characte ristics Abrasion resistance	Non-magnetic steel
	Mg Forming	PC chassis Mobile phone chassis	Durability Light weight	Heat resistance Abrasion resistance	Non-magnetic steel

	Plastic resin	Example of molded product	Characte-ristics expected of the plastic resin	Characte-ristics expected of the steel	Recommend-ed type of steel
Thermoset Resin	Phenol resin Melamine resin	Kitchen utensils Ash trays	Heat resistance	Abrasion resistance	Pre-hardened steel SKD11
	Phenol resin Un-saturated polyester	Switches Connectors	Heat resistance Flaming resistance	Abrasion resistance Corrosion resistance	SUS420J2 SKD11
	Epoxy resin	IC sealing Transistors	Electrical insulation	Abrasion resistance Corrosion resistance	SUS420J2 SKD11

E 006 Method of Preliminary Drying of Plastic Resin

Pellets of plastic resin generally absorb the moisture in the air to some extent.

If the amount of moisture absorbed is high, the plastic resin can undergo hydrolysis in the process of being melted and kneaded inside the cylinder of the injection molding machine (there are also plastics that undergo chemical dissociation with water as the trigger), or silver-colored lines (silver streaks) appear on the surface of the molded item at the time of injection molding, or defects such as air bubbles, insufficient glossiness, transfer defects, etc. can occur.

In view of this, it is necessary that moisture in the pellets of the molding material is removed beforehand by putting them through a drying unit.

If this preliminary drying is not done properly, it can cause variations in the fluidity or reduction in the material properties, and molding defects.

The following are the main types of drying units used for this purpose.

1. Hot air dryer

The hopper dryer and the box type drying oven are typical of this type dryer. This is a method in which the moisture content is evaporated by blowing hot air on the pellets. Although this is very common and simple drying method, it is not suitable for thoroughly removing the moisture content.

E-006 플라스틱 수지의 예비 건조의 방법

플라스틱 수지의 펠릿은 일반적으로 공기 중의 수분을 어느 정도 흡수한다.

흡습량이 많으면 사출성형기의 실린더 내부에서 용융 혼련하는 과정에서 플라스틱 수지가 가수 분해될 수 있다. 사출성형 할 때에 성형품의 표면에 은색 선(실버 스트릭)이 나타나거나, 기포, 불충분한 광택, 전사 결함 등이 발생한다(예를 들어, 물과의 화학적 해리를 일으키는 플라스틱).

이러한 관점에서 성형 재료의 펠릿 내의 수분은 건조기를 통과시켜 미리 제거 할 필요가 있다.

예비 건조가 적절하게 이루어 지지 않으면, 유동성 변화 또는 재료 특성의 감소, 성형 결함에 영향을 미친다.

다음은 이러한 목적으로 사용되는 건조기의 주요 유형이다.

1. 열풍건조

호퍼 드라이어 및 박스형 건조 오븐이 이러한 유형의 건조기의 전형적인 예이다. 이것은 펠릿에 뜨거운 공기를 불어 넣어 수분 함량을 증발시키는 방법이다. 매우 보편적이며 간단한 건조 방법이지만 수분 함량을 완전히 제거하는 데는 적합하지 않다.

2. Dehumidified hot air dryer

This is the method in which the moisture content in the pellets is evaporated by blowing a hot air the moisture content in which has been removed in advance.
Since the hot air used for drying can be used again after dehumidifying it again, heat loss is low, and it is possible to carry out rational drying.

This method is suitable for drying PBT, etc

2. 제습 열풍 건조기

제습 열풍 건조기는 사전에 수분을 제거한 뜨거운 공기로 불어 넣어 펠릿 내의 수분 함량을 증발시키는 방법이다. 건조에 사용 된 뜨거운 공기는 다시 제습 한 후에 다시 사용할 수 있기 때문에 열 손실이 적으므로 합리적인 건조가 가능하다.

이 방법은 PBT 건조 등에 적합하다.

3. Reduced pressure heat transfer type dryer

This is the method of evaporating the moisture content in the pellets by heat transfer in a reduced pressure environment. Drying becomes possible at a low temperature, and it is possible to prevent oxidization of the plastic and to reduce the effect of additives inside the pellet.
In addition, this is also a method in which heat loss is small.
This type of dryer is attracting a lot of attention as the drying unit of the future.

* Reference: "An Encyclopedia of Injection Molding", 'Peripheral and Auxiliary Equipment', p. 214, (Hideki Kubo, Industrial Research Center of Japan, (2002)) (in Japanese)

3. 감압 전열식 건조기

감압 전열식 건조기는 감압 환경에서의 열전달에 의해 펠릿 내의 수분을 증발시키는 방법이다. 저온에서 건조가 가능하고, 플라스틱의 산화를 방지하고, 펠릿 내부의 첨가제의 효과를 감소시킬 수 있다.

또한 열 손실이 적은 방법이기도 하다.

이 건조기는 미래의 건조 장치로서 많은 주목을 받고 있다.

E 007 Biodegradable Plastics (1)

A biodegradable plastic is a material that, when buried in underground, etc., gets dissociated biologically due to bacteria or microorganisms.

This characteristic of such plastics can be considered to be ideal from the point of view of protecting the environment as compared to plastics based on petroleum.

The development of biodegradable plastic materials having the properties needed in the industry such as transparency, heat resistance, etc., is progressing rapidly. These are being more and more widely used in electrical home appliances, components for OA equipment, food containers, interior decoration parts of automobiles, etc.

In the Aichi Expo held in March 2005 (Aichi World Expo), biodegradable plastics were used in various ways. For example, the food utensils used in the restaurants were all made of biodegradable plastics. Large PC manufacturers have also announced that they will be using biodegradable plastics for the components in a PC, and the demand for biodegradable plastics is considered to increase rapidly from now on in view of the laws on environmental protection and recycling.

In the field of automobile interior decoration parts, Toyota Motor Company has announced that they will be using poly-lactic acid products in successive stages.

The following materials are the major biodegradable plastics. The manufacturers of the material are given within parentheses ( ).

E-007 생분해성 플라스틱 (1)

생분해성 플라스틱은 매장되었을 때 박테리아나 미생물로 인해 생물학적으로 분해되는 물질이다.

이러한 플라스틱의 특성은 석유를 기반으로 하는 플라스틱에 비하여 환경 보호의 관점에서 이상적이다.

투명성, 내열성 등 업계에서 요구되는 특성을 갖는 생분해성 플라스틱 소재의 개발이 급속히 진행되고 있다. 이들은 가전제품, OA 기기용 부품, 식품 용기, 자동차 내장 부품 등에 점점 더 널리 사용되고 있다.

2005년 3월에 개최 된 아이치 엑스포(Aichi Expo)에는 생분해성 플라스틱이 여러 면에서 소개되었다. 예를 들어 레스토랑에 사용된 식기류는 모두 생분해성 플라스틱으로 만들어졌다. 대형 PC 제조업체들은 PC의 구성 요소에 생분해성 플라스틱을 사용할 것이라고 발표했으며, 생분해성 플라스틱에 대한 수요는 환경 보호 및 재활용에 관한 법률에서 볼 때 급속도로 증가하는 것으로 생각된다.

Toyota Motor Company는 자동차 내장 부품 분야에서 폴리 락틱산 제품을 연속적으로 사용할 것이라고 발표했다.

다음 재료는 주요 생분해성 플라스틱이다. 재료의 제조업체는 괄호 안에 나타내었다.

- Poly-lactic acid (Nature Works, Mitsui Chemicals, Unitika, Toyobo, Dai Nippon Ink Chemicals, Toyota Motor Company, and others)
- Poly-butylene succinate (Mitsubishi Chemicals, Showa Polymers)
- Poly-ethylene succinate (Nippon Catalysts)
- Polyvinyl alcohol (Kuraray, Nihon Synthetic Chemical Industries)
- Poly-glycolic acid (Kureha Chemicals)
- Cellulose acetate (Daicel Chemical Industries)

E 008 Biodegradable Plastics (2)

E-008 생분해성 플라스틱 (2)

Biodegradable plastic is a material that gets dissociated biologically due to bacteria or microorganisms either underground or in air.

The conventional plastic resin synthesized by petroleum chemistry cannot be dissociated by bacteria, and are considered to remain permanently as waste. However, due to the recent advances in chemical technology, plastic resin materials have been developed that are dissociated into water and carbon dioxide. Some of such materials have already been made into commercial products and are being used in various injection molded items.

After using the molded items produced from a biodegradable plastic resin, they are buried underground,

생분해성 플라스틱은 땅속 또는 대기 중의 박테리아 또는 미생물로 인해 생물학적으로 분해되는 물질이다.

석유 화학으로 합성된 기존의 플라스틱 수지는 박테리아에 의해 분해 될 수 없으며, 폐기물로 영구적으로 남는다. 그러나 최근 화학 기술의 발전으로 인하여 물과 이산화탄소로 분해되는 플라스틱 수지 재료가 개발되었다. 이러한 재료 중 일부는 이미 시판 제품으로 만들어졌으며 다양한 사출성형품에 사용되고 있다.

생분해성 플라스틱 수지로 만든 성형품을 사용하면 땅속에 묻히거나

or are mixed with kitchen garbage as compost and left to decompose. These are ideal materials for preserving the earth's environment and for preventing global warming.

In order to use biodegradable plastics resin in industrial products or as packaging materials, while it is necessary that they have a certain strength and heat resistance, and satisfy the standards related to food and hygiene, products satisfying these requirements are being successfully commercialized one after the other.

The tableware used in the restaurant in the Aichi World Expo held in March of 2005 were made of biodegradable plastics.

It is considered that the practical realization of biodegradable plastics resin will progress at a very fast pace in the fields of electrical home appliances, interior decoration parts of automobiles, tableware etc, and in various other fields.

Some of the major biodegradable plastic materials are described below.

퇴비로 부엌 쓰레기와 섞여 분해된다. 이것은 지구의 환경을 보존하고 지구 온난화를 방지하는 데 이상적인 물질이다.

생분해성 플라스틱 수지를 공업제품 또는 포장재로 사용하기 위해서 일정한 강도와 내열성을 갖고 식품 및 위생 관련 기준을 충족시켜야하지만 이러한 요구 사항을 만족하는 제품이 하나씩 상용화되고 있다.

2005년 3월에 열린 아이치 엑스포에서 식당에서 사용된 식기류는 생분해성 플라스틱으로 만들어졌다.

가전제품, 자동차 내장재, 식기류 등 다양한 분야에서 생분해성 플라스틱 수지의 실용화가 매우 빠른 속도로 진행될 것으로 생각된다.

주요 생분해성 플라스틱 재료 중 일부는 아래에 설명되어 있다.

1. Poly-lactic acid (PLA)

This is a plastic resin that is synthesized only from starch produced from corn or potatoes, and is a material whose commercialization has progressed most.

Grades have appeared that can improve the heat resistance and strength, and are being used practically in parts of electrical home appliances, computer products, returnable food containers (bowls, cups, etc. that can be used in dish washers).

In the case of heat resistant grades, although problems have been pointed out that it is necessary to maintain the mold temperature at 110 to 120℃, and

1. 폴리-락틱산 (PLA)

폴리-락틱산은 옥수수 또는 감자 전분으로부터 제조되는 합성 플라스틱 수지이며, 대부분이 상업화에 사용되는 재료이다.

내열성과 강도를 향상시킨 그레이드가 등장 해 가전제품, 컴퓨터 제품, 반환 식품 용기(식기 세척기에 사용할 수있는 그릇, 컵 등)의 일부에 실제로 사용되고 있다.

내열 등급의 경우 금형 온도를 110 ~ 120℃로 유지할 필요가 있고,

that separating from the mold is very difficult, production methods have been developed to solve these problems.

금형에서 분리하는 것이 매우 어렵다는 문제가 지적되었지만 이러한 문제를 해결하기 위한 생산 방법이 개발되었다.

2. Poly butylene succinate (PBS)

This is a biodegradable plastic resin made from soft biomass of which succinic acid is a typical constituent. While poly-lactic acid is suitable for high hardness molded items, PBS is suitable for soft molded items.

2. 폴리 부틸렌 숙시네이트 (PBS)

폴리 부틸렌 숙시네이트는 전형적인 숙신산이 성분인 부드러운 바이오메스로 만든 생분해성 플라스틱 수지이다. 폴리 락틱산은 고경도 성형 제품에 적합하지만 PBS는 소프트 성형 제품에 적합하다.

3. Poly vinyl alcohol (PVA)

3. 폴리 비닐 알코올 (PVA)

4. Poly glycolic acid

4. 폴리 글리콜 산

E 009 Glass Transition Temperature Tg of Plastics

E-009 플라스틱의 유리 전이 온도 Tg

Thermoplastic resins melt when its temperature becomes higher than its melting point and becomes liquid, but as it is gradually cooled from the liquid state its viscosity gradually increases, and it goes into a rubbery state and it finally solidifies. The temperature at the border from the rubbery state to the solid state (called the glass state) is called its "glass transition temperature". The glass transition temperature is generally indicated as Tg.

열가소성 수지는 용융점보다 높은 온도에서 용융되어 액상이 되지만 액체 상태에서 서서히 냉각됨에 따라 점노가 점차 증가하여 고무 상태가 되어 최종적으로 응고된다. 고무 상태에서 고체 상태로의 온도(유리 상태라고 함)를 "유리 전이 온도"라고 한다. 유리 전이 온도는 일반적으로 Tg로 표시된다.

Tg is a physical quantity that is determined by the type of plastic. By knowing the value of Tg, it is possible to estimate the time required for the molded item to cool inside the cavity and the molded item temperature at which it can be ejected from the mold.

Tg는 플라스틱의 유형에 따라 결정되는 물리적인 양이다. Tg의 값을 알면 캐비티 내부에서 성형품이 냉각되는 데 필요한 시간과 금형에서 취출 될 수 있는 성형품 온도를 예측할 수 있다.

Name of plastic	Tg	Name of plastic	Tg	Name of plastic	Tg	Name of plastic	Tg
Polyethylene	-125℃	Polyurethane	-20℃	Polyamide 6	50℃	EVA	-42℃
Polypropylene	0℃	Polyether sulfone	230℃	Polyamide 66	50℃	Polyacrylonitrile	104℃
Polyvinyl chloride	87℃	Polyphenylene oxide	120℃	Polyacetal	-50℃	Polyfluoro vinylidene	35℃
Polystyrene	100℃	Polyamide imide	275℃	Polycarbonate	150℃	Polypropylene terephthalate	50℃
ABS	125℃	Polylactic acid	57℃	Polyphenylene sulfide	126℃	Polyethylene terephthalate	69℃
Polymethyl methacrylate	90℃	Polytetrafluoroethylene	126℃	-	-	-	-

E 010 Melt Flow Rate (MFR) of Plastics

E-010 플라스틱 수지의 용융 흐름 지수

In the injection molding of plastics resin, you may all have experienced that the settings of the molding conditions, particularly of the injection pressure or the injection speed, and of the mold temperature are greatly affected by whether or not the fluidity of the molten plastic is good or bad. While there are several methods for evaluating the fluidity of plastics, the easiest method which is being used as a guideline is the Melt Flow Rate (MFR) method.

플라스틱 수지의 유동성이 양호하거나 불량하든 상관없이 사출성형 조건, 특히 사출압력 또는 사출속도 및 금형온도의 설정이 용융수지의 유동성에 큰 영향을 끼친다. 플라스틱의 유동성을 평가하는 여러 가지 방법이 있지만 지침으로 사용되는 가장 쉬운 방법은 용융 흐름 지수(MFR) 방법이다.

The melt flow rate is measured by putting the plastic resin under test in a testing equipment called a melt low indexer, heating the plastic resin, making the molten plastic resin with a prescribed weight to flow, and measuring the weight of the plastic that flows out as an index of the melt flow rate. The method of this test has been stipulated in JIS and ISO standards. A plastic resin with a high value of melt flow rate is evaluated as one with good fluidity. The flow is poor of a plastic resin whose melt flow rate is low. However, the melt flow rate evaluates the fluidity in a static condition, whereas in actual injection molding the plastic resin flows in a very short time through a very narrow gate, so it is necessary to understand well the fact that the melt flow rate as it is does not fit in the case of injection molding.

As explained above, the melt flow rate is used as an index of the fluidity of the plastic resin for grasping the properties of the plastic resin in a simple manner.

용융 흐름 지수는 멜트 로 인덱서라고 불리는 시험 장치에 플라스틱 수지를 투입하여 가열하고, 소정 중량의 용융 수지를 유동시키고, 유출 된 플라스틱의 중량을 측정한다. 용융 흐름 지수는 유속의 지표로서 사용한다. 이 시험 방법은 JIS 및 ISO 규격에 규정되어 있다. 용융 흐름 지수가 높은 플라스틱 수지는 유동성이 좋은 것으로 평가된다. 용융 흐름 지수가 낮은 플라스틱 수지의 유동성은 낮다. 그러나 용융 흐름 지수는 정적 상태에서의 유동성을 평가하는 반면, 실제 사출성형에서 플라스틱 수지는 아주 짧은 시간에 매우 좁은 게이트를 통해 흐르기 때문에 용융 흐름 지수가 사출성형의 경우에는 맞지 않는다는 사실을 잘 이해할 필요가 있다.

위에 기술한 바와 같이, 용융 흐름 지수는 플라스틱 수지의 물성을 간단히 파악하기 위해서 플라스틱 수지의 유동성의 지표로서 사용된다.

Name of plastic resin	Melt flow rate	Name of plastic resin	Melt flow rate
Polypropylene	5 to 60	Polycarbonate	2 to 30
Polystyrene	5 to 25	PBT	20 to 100
AS resin	10 to 40	PPO	7 to 30
ABS resin	5 to 60	PSU	7 to 18
Acrylic resin	1 to 35	PES	17 to 30
Polyamide 6	10 to 80	Cyclo olefin polymer	7 to 60
Polyamide 66	10 to 80	TPX	21 to 80
Polyacetal	3 to 70	PEEK	1 to 20

E 011 Plant-Origin Biodegradable Plastic - Polylactic Acid Plastic resin (1)

Polylactic Acid (PLA) is thermoplastic plastic using 100% plant-origin raw materials, and lends itself easily to injection molding, extrusion molding, and blow molding. In addition, it is a plastic resin that can be synthesized without using any petroleum based or chemical based raw materials. Furthermore, it has the excellent environmental characteristic that when this plastic is disposed of in soil, it is completely decomposed into water (H_2O) and carbon dioxide (CO_2) by the enzymes produced by the bacteria in the soil.

Although the presence of this plastic was confirmed more than ten years ago, the method of manufacturing it industrially on a large scale had not been established, and it was almost never specified for use in mass produced items. However, Cargill Dow LLC, which is a joint venture between Cargill Inc. and Dow Chemicals Inc. of USA, succeeded in developing an industrial method of mass manufacturing lactic acid, and it started to be sold in the market. At present, this company has changed its name to NatureWorks LLC and is expanding its business.

The raw material for the polylactic acid plastic is of vegetable origin as mentioned above. At present, starch or sugar (glucose) is being used. In other words, the raw materials will be corn or various tubers, sugarcane, sugar beets, etc. When using starch as the raw material, a reaction called hydrolysis occurs when water is added to starch, it chemically changes to glucose. This is the same reaction as that which occurs when we chew rice and it gradually tastes sweet.

E-011 식물원료 생분해성 플라스틱 - 폴리락틱산 플라스틱 수지(1)

폴리락틱산(PLA)은 식물성 원재료를 100% 사용하는 열가소성 플라스틱으로 사출성형, 압출성형 및 블로우 성형에 쉽게 사용된다. 또한 석유계 또는 화학 원료를 사용하지 않고 합성 할 수 있는 플라스틱 수지이다. 또한 이 플라스틱을 토양에 폐기하면 토양의 박테리아가 생산하는 효소에 의해 완전히 물(H_2O)과 이산화탄소(CO_2)로 분해되는 뛰어난 환경적 특성을 가지고 있다.

이 플라스틱의 존재가 10년 이상 전에 확인되었지만, 산업적으로 대규모로 제조하는 방법은 확립되지 않았으며, 대량 생산품에 사용되는 것은 거의 알려지지 않았다. 그러나 Cargill 사와 미국 Dow Chemicals 사의 조인트 벤처인 Cargill Dow LLC는 젖산 생산의 산업적 방법을 개발하는데 성공하여 시장에서 판매되기 시작했다. 현재 이 회사는 NatureWorks LLC로 이름을 바꾸고 사업을 확장하고 있다.

폴리락틱산 플라스틱 원료는 위에서 설명한 바와 같이 식물성 원료이다. 현재 전분이나 설탕(포도당)이 사용되고 있다. 즉, 원료는 옥수수 또는 여러 가지 덩이줄기, 사탕수수, 사탕무 등이 될 것이다. 전분을 원료로 사용하면 전분에 물을 첨가하면 화학적으로 포도당으로 변한다. 이것은 우리가 찹쌀을 씹을 때와 같은 반응이며 서서히 단맛이 난다.

Next, lactic acid bacteria (lactobacillus) are added to the glucose. These microorganisms called lactic acid bacteria have the ability to chemically convert glucose into lactic acid. Lactic acid is a material that is synthesized naturally inside the human body, and it has been known to cause muscle pain such as neck pain or shoulder pain when it accumulates in the muscles.

Next, by carrying out a treatment called dehydration reaction of this lactic acid, it is possible to change it into a material called lactide. There are two types of lactide, namely, L-lactide and D-lactide, but mostly it will be L-lactide. Although whether the result of reaction is L-lactide or D-lactide depends on the type of lactic acid bacteria, not many bacteria have been found that can be used for manufacturing D-lactide.

Next, by carrying out processing called ring opening polymerization of the above lactide, it is possible to obtain polylactic acid plastic.

Polylactic acid is a polymer and is thermoplastic. Therefore, it melts when heated to a high temperature and becomes a solid when cooled. The glass transition temperature is around 57℃. In other words, it is possible to mass produce items by injection molding using molds.

Injection molding of normal polylactic acid plastic can be made by setting the mold temperature to about 30℃ to 40℃. Even the design of molds can be made once some experience has been gained. However, since the withstanding temperature of the molded items will be around 60℃, and since even their mechanical strength is also not high, the applications of the molded items will be limited to those satisfying these conditions. Transparent grades are also available. As a result of research in various companies, polylactic acid plastics

다음은 락토바실리스(lactobacillus)이 포도당에 첨가된다. 유산균이라고 불리는 이런 미생물은 포도당을 젖산으로 화학적으로 전환시키는 능력을 가지고 있다. 락틱산은 인체 내부에서 자연적으로 합성되는 물질로 근육에 축적되면 목통증이나 어깨통증과 같은 근육통을 유발하는 것으로 알려져 있다.

다음은 이 락틱산 탈수 반응이라 불리는 처리를 하여 락티드라고 하는 물질로 변환 할 수 있다. 락티드에는 L-락티드와 D-락티드의 두 가지 유형이 있지만 대부분 L-락티드이다. 반응 결과가 L-락티드인지 D-락티드인지는 유산균의 종류에 따라 다르나, D-락티드의 제조에 사용할 수 있는 박테리아는 많지 않다.

다음은 상기 락티드의 개환 중합이라 불리는 처리를 하므로 폴리락틱산계 플라스틱을 얻을 수 있다.

폴리락틱산은 폴리머이며 열가소성이다. 따라서 고온으로 가열하면 용융되고, 냉각되면 고체가 된다. 유리 전이 온도는 약 57℃ 이다. 즉, 금형을 이용하여 사출성형하면 대량 생산이 가능하다.

금형 온도를 약 30 ~ 40℃로 설정하여 일반 폴리락틱산 플라스틱의 사출성형이 가능하다. 경험이 축적되면 금형설계도 가능하다. 그러나 성형품의 내열 온도는 약 60℃이기 때문에 기계적 강도가 높지 않고, 성형품의 용도는 이들 조건을 만족하는 것에 한정된다. 투명한 그레이드도 가능하다. 여러 회사의 연구 결과로 폴리락틱산 플라스틱은 현재 약 120℃의 온도를 견딜 수 있게

are now being sold that can withstand temperatures of around 120℃. In the case of these grades, care should be taken during the mold design because the fluidity is bad, mold releasing is extremely bad, and also the cooling time becomes long. The scope of applications of polylactic acid plastics has become considerably wide due to the advent of the high heat resistant grade. Studies are now under way in various fields about the application of these plastics.

판매되고 있다. 이 등급의 경우 유동성, 이형성이 매우 나쁘고, 냉각 시간이 길어지므로 금형설계 시에 주의를 기울여야한다. 폴리락틱산 플라스틱의 적용 범위는 높은 내열 그레이드의 출현으로 인해 상당히 넓어졌다. 현재 이 플라스틱의 적용에 대한 다양한 분야의 연구가 진행 중이다.

E 012 Plant-Origin Biodegradable Plastic - Polylactic Acid Plastic (2)

E-012 식물원료 생분해성 플라스틱 - 폴리 락틱산 플라스틱(2)

Polylactic Acid (PLA) is a plastic resin that can be manufactured using only plant-origin raw materials without using any petroleum based or chemical based raw materials, and in addition, it is an ideal environment-friendly plastic that is decomposed completely into water and carbon dioxide by the actions of naturally occurring bacteria after it has been disposed of. However, the following doubts come to mind. Since appropriate information is difficult to obtain regarding these doubts, some people may come to misunderstandings about them. The points of view that are supported by a large number of concerned persons at present are described below.

Polylactic Acid (PLA)는 석유를 원료로하거나 화학 물질을 사용하지 않고 식물성 원재료만을 사용하여 제조 할 수 있는 플라스틱 수지이며, 또한 폐기 된 후에 자연적으로 발생하는 박테리아의 작용에 의해 물과 이산화탄소로 완전히 분해되는 이상적인 환경 친화적 플라스틱이다. 그러나 다음과 같은 의심이 떠오른다. 이러한 의문에 대해 적절한 정보를 얻기 어렵기 때문에 일부 사람들은 그것에 대한 오해를 하게 된다. 현재 다수의 관계자가 지지하는 견해가 아래에 설명되어 있다.

Q1. Why is that polylactic acid resin is not in widespread use in spite of being an excellent material?

Q1. 폴리락틱산 수지가 우수한 소재 임에도 불구하고 널리 사용되지 않는 이유는?

▶ **A1.**

While many reasons can be thought of, the most important reason is that the cost of the raw material is still high compared to the raw materials of polyolefin plastics (polypropylene, polyethylene, etc.) or styrene based plastics (PS, ABS, etc.). It was inevitable that the cost of raw materials was still high because the production volume is still low. However, the cost difference is gradually becoming smaller because of big increases in the cost of petroleum and because of the sudden increase in the production volume of polylactic acid plastics. As a result, the trend is towards the problem of the cost of raw materials becoming smaller.

▶ **A1.**

많은 이유가 생각 될 수 있지만, 가장 중요한 이유는 폴리올레핀 플라스틱 (폴리프로필렌, 폴리에틸렌 등) 또는 스티렌 기반 플라스틱 (PS, ABS 등)의 원료에 비하여 비용이 여전히 높다는 것이다. 생산량이 여전히 낮기 때문에 원자재 가격이 여전히 높았던 것은 불가피했다. 그러나 석유 가격이 크게 상승하고, 폴리락틱산 플라스틱의 생산량이 갑자기 증가함에 따라 비용 차이가 점점 작아지고 있다. 결과적으로 원재료 비용이 점점 낮아지는 추세이다.

Q2. Is it not possible to burn off polylactic acid ?

▶ **A2.**

It can also be burnt off similar to fossil fuel based plastics. The amount of carbon dioxide and heat generated at that time is about half that of fossil fuel based plastics, and it has been confirmed that these plastics are superior even when burnt in terms of the effect on global warming.

Q2. 폴리락틱산을 태울 수 없습니까?

▶ **A2.**

화석 연료 기반의 플라스틱과 유사하게 연소 될 수 있다. 연소 시 발생 된 이산화탄소와 열량은 화석 연료 기반 플라스틱의 약 절반으로 지구 온난화의 영향과 관해서도 연소 시에도 이 플라스틱이 우수하다는 것이 확인되었다.

Q3. Is it not possible to reuse the runner, etc. in the case of polylactic acid?

▶ **A3.**

They can be reused.

Q3. 폴리락틱산의 경우 런너 등의 스크랩을 재사용 하는 것이 불가능 한가?

▶ **A3.**

다시 사용할 수 있다.

Q4. What time period is required for biodegradation?

A4.

For biodegradation to start, it is necessary that the environment is suitable, such as the quantity of microorganisms, temperature, humidity, pH, etc. Biodegradation does not start under the normal environment of use in our daily life (such as stationery, container, etc.) just because of touching by hand or by keeping water at room temperature in a container. Biodegradation starts in soil or in compost if the conditions are satisfied. In soil, about one month is required for biodegradation to start. In compost, biodegradation starts after about three days if the temperature inside the compost increases to near 60℃. A film shaped molded item put in compost gets completely decomposed in about two weeks. If a molded item with a wall thickness of about 1 ㎜ is buried in soil, it takes about four to five years for it to get completely decomposed.

Q4. 생분해를 위해서 필요한 기간은?

A4.

생분해가 시작되기 위해서 미생물의 양, 온도, 습도, pH 등과 같은 환경이 적절해야 한다. 일상생활에서 정상적인 환경에서는 생분해가 시작되지 않는데(예 : 문구류, 용기 등) 왜냐하면 단지 손으로 만지거나 용기에 실온의 물을 담아 두기 때문이다. 조건이 만족되면 토양 또는 퇴비에서 생분해가 시작된다. 토양에는 생분해가 시작되기까지 약 1개월이 필요하다. 퇴비에서 온도가 60°C 가까이 상승하면 약 3일 후에 생분해가 시작된다. 퇴비에 넣은 필름 모양의 성형품은 약 2주 만에 완전히 분해된다. 살 두께가 약 1 ㎜인 성형품을 토양에 묻으면 완전히 분해되는 데 약 4 ~ 5년이 걸린다.

Q5. Is there no danger of causing a food shortage problem if corn is used as the raw material?

A5.

The corn that is being used at present for the manufacture of polylactic acid plastics is millet that is being produced for use as animal fodder for cows, horses, etc. It is not the sweet corn eaten by people. About 99% of the corn being produced in the world is for fodder, and the corn consumed directly as food by people is estimated to be about 1%. At present, the production volume of polylactic acid is about 0.01% of the world's plastic production including those based on petroleum, and even if it increases to about 10%, it is

Q5. 옥수수가 원료로 사용될 경우 식량 문제를 일으킬 위험이 없는가?

A5.

현재 폴리락틱산 플라스틱 제조에 사용되는 옥수수는 소, 말 등의 동물 사료로 사용하기 위해 생산된다. 사람들이 먹는 단 옥수수는 아니다. 세계에서 생산되는 옥수수의 약 99%가 사료용이며 옥수수를 직접 먹는 사람들은 약 1%로 추정된다. 현재 폴리락틱산의 생산량은 석유를 비롯한 세계 플라스틱 생산량의 약 0.01%이며, 약 10%까지 증가하더라도 수수의 생산 능력을 고

almost impossible to think of this corn usage ever becoming a problem of food for human consumption considering the production capacity of millet.

려하면 인간의 소비를 위한 식량 문제가 될 것이라고 생각하지는 않는다.

Q6. Can polylactic acid plastics be eaten?

Q6. 폴리락틱산 플라스틱을 섭취 할 수 있습니까?

▶ A6.

Polylactic acid plastic is not edible. However, it is harmless even if consumed accidentally. When using for food containers, it is necessary to use a grade that conforms to safety standards of the industry and laws and regulations.

폴리락틱산 플라스틱은 먹을 수 없다. 그러나 우발적으로 먹게 되더라도 무해하다. 식품 용기에 사용하는 경우 산업 안전 기준 및 법령을 준수하는 등급을 적용해야 한다.

Q7. Is it possible to add coloring agents to polylactic acid plastics?

Q7. 폴리락틱산 플라스틱에 착색제를 첨가 할 수 있는가?

▶ A7.

Coloring is possible in master batches, etc. Coloring agents that are biodegradable are also being developed.

마스터 배치 등에서도 착색이 가능하다. 생분해성인 착색제도 개발되고 있다.

E 013 Thermal Diffusion Coefficient of Major Plastics

E-013 주요 플라스틱의 열 확산 계수

In order to calculate the cooling time of plastic injection molded items, it is necessary to acquire the thermal diffusion coefficient of the plastic resin. Table 1 gives the thermal diffusion coefficients of some major plastic resins.

The thermal diffusion coefficient of the plastic at the cavity surface temperature (㎟/sec)

플라스틱 사출성형품의 냉각시간을 계산하려면 플라스틱 수지의 열 확산 계수를 알아야한다. 표 1은 일부 주요 플라스틱 수지의 열확산 계수를 나타낸다.

캐비티 표면 온도에서의 플라스틱의 열 확산 계수 (㎟/sec)

$$\alpha = \lambda / (c \cdot \rho)$$

λ : Thermal conductivity of the plastic resin(kcal/m • h • ℃)

c : Specific heat of the plastic resin(kcal/kg • ℃)

ρ : Density of the plastic resin(kg/m^3)

$$\alpha = \lambda / (c \cdot \rho)$$

λ : 플라스틱 수지의 열전도율 (kcal/m • h • ℃)

c : 플라스틱 수지의 비열 (kcal/kg • ℃)

ρ : 플라스틱 수지의 밀도(kg/m^3)

Table 1 Thermal diffusion coefficient of major plastic resins

표 1 주요 플라스틱 수지의 열확산계수

Cavity surface temperature	Plastic material						
	PS	ABS	PP	PMMA	PA66	POM	PC
30℃	0.0851	0.0818	0.0664	0.0766	0.0885	0.0762	0.1074
40℃	0.0832	0.0822	0.0653	0.0750	0.0880	0.0724	0.1078
50℃	0.0813	0.0827	0.0643	0.0734	0.0875	0.0685	0.1083
60℃	0.0794	0.0831	0.0632	0.0718	0.0870	0.0647	0.1087
70℃	0.0774	0.0835	0.0621	0.0702	0.0865	0.0609	0.1092
80℃	0.0755	-	0.0611	0.0686	0.0860	0.0571	0.1096

E 014 PET Plastic

PET is the abbreviation for polyethylene terephthalate plastic resin.

PET has a melting point near 270℃, while the molding temperature is 270 to 280℃, and hence its injection molding is done at a relatively high temperature. One of the features of this plastic is its fluidity. Although its fluidity is good above the melting point, the flow suddenly becomes difficult when the solidification starts. In other words, there is a very close

E-014 PET 플라스틱

PET는 폴리에틸렌 테레프탈레이트 플라스틱 수지의 약자이다.

PET는 270°C 근처의 융점을 가지며 성형 온도는 270 ~ 280°C이므로 사출성형은 상대적으로 높은 온도에서 수행된다. 이 플라스틱의 특징 중 하나는 유동성이다. 유동성이 융점 이상에는 우수하지만 고화가 시작되면 유동이 갑자기 어려워진다. 즉, 플라스틱의 온도와 유동성

relationship between the temperature of the plastic and the fluidity. Temperature management of the mold and temperature management of the plastic are mandatory items in the case of PET.

In most cases, hot runners are used in the injection molding of PET, and in addition, valve gates are selected in order to make certain the management of the fluidity by mechanically opening and closing the gate.

Since the fluidity is good, it is necessary to give considerations to the mold clearance management and to mold deformations due to the filling pressure. In particular, in the design of the manifold of the hot runner, the key points will be prevention of deformation and management of the thermal expansion margin.

Further, since PET reacts sensitively to water content, it is necessary to carry out thorough preliminary drying of the molding material pellets and also to make the management level high. This plastic carries out hydrolysis when it reacts with water. Therefore, the material pre-drying equipment is mandatorily required in the injection molding of PET.

In the process of solidifying from the fluid state, the size of the crystals of PET becomes larger and the crystals can easily become spherocrystals. When this happens the molded articles will not be transparent but will become turbid white in color. Consequently, if transparent molded items are to be prepared, it is necessary to cool rapidly in order to suppress the growth of spherocrystals. Therefore, the molds should be provided with a system for rapid cooling. It is necessary to maintain the mold temperature between 5 and 15℃, for which a chiller is used. Since it is necessary to manage the region of change of plastic temperature inside the mold in the temperature gradient from 270 to 280℃ to 5 to 15℃, it can be understood how difficult

사이에는 매우 밀접한 관계가 있다. 금형의 온도 관리 및 플라스틱의 온도 관리는 PET의 경우 필수 항목이다.

대부분의 경우, PET의 사출성형에 핫 런너가 사용되며, 게이트를 기계적으로 열고 닫음으로써 유동성 관리를 확실하게하기 위해 밸브 게이트가 선택된다.

유동성이 양호하므로 금형 틈새 관리 및 충전압력으로 인한 변형을 고려해야한다. 특히 핫 런너의 매니폴드 설계 시 열팽창으로 인한 변형 및 관리를 방지하는 것이 중요하다.

또한, PET는 수분에 민감하게 반응하기 때문에 펠릿의 예비 건조를 철저히 할 필요가 있고, 관리 레벨을 높게 할 필요가 있다. 이 플라스틱은 물과 반응 할 때 가수 분해를 한다. 따라서 PET의 사출성형에는 사전 건조 장비가 필수적으로 요구된다.

유체 상태에서 고화하는 과정 중에 PET의 결정 크기가 커지고, 결정이 쉽게 구형 결정이 될 수 있다. 이런 일이 발생하면 성형품은 투명하지는 않고, 색상이 탁하며, 희게 된다. 결과적으로, 투명한 성형품을 제조하는 경우, 구형 결정의 성장을 억제하기 위해 신속하게 냉각 할 필요가 있다. 따라서 금형에 급속 냉각 시스템을 제공해야 한다. 냉각기를 사용하는 금형 온도 변화를 5 ~ 15°C로 유지해야 한다. 금형 내에서 플라스틱 수지의 온도가 270 ~ 280℃에서 5 ~ 15℃의 온도로 관리해야 하므로 PET의 사출성형이

the injection molding of PET is.

At present, the injection molding technology of PET plastic has advanced and reached a mature stage, and the mold technology of super multiple shot molds (100 to 200 shots) is being used in the world in the mass production mode.

Although the raw material of PET plastic is a petrochemical material, at present bio-PET plastic is being developed. This makes the raw material to be produced from vegetable origin, and it is possible to produce a plastic with completely the same composition as the petrochemical based PET plastic. Supply of such material on a commercial basis has already started.

The locations supplying the raw materials for PET plastic is predicted to be not oil producing countries but is predicted to shift gradually towards agricultural products producing countries.

얼마나 어려운지 알 수 있다.

현재 PET 플라스틱의 사출성형 기술은 성숙 단계에 이르렀고, 대량 생산이 가능하며, 대량생산 세계에서 수퍼 다중 사출 금형 (100 ~ 200샷) 기술이 사용되고 있다.

PET 플라스틱의 원료는 석유 화학 소재이지만 현재 바이오 PET 플라스틱이 개발 중이다. 이로 인해 원료가 식물성 원료에서 생산되며, 석유 화학 기반 PET 플라스틱과 완전히 동일한 조성의 플라스틱을 생산할 수 있다. 이러한 물질의 상업적 공급은 이미 시작되었다.

PET 플라스틱 원료를 공급하는 곳은 석유 생산국이 아니라 농산물 생산 국가로 점차 옮겨 갈 것으로 예상된다.

E 015 Injection Molding of Heat Resistant Polylactic Acid Resins

One of the typical biodegradable plastics is polylactic acid (PLA). Polylactic acid is a plant-derived biodegradable plastic resin. It is chemically synthesized using the following synthesizing process, and is recycled using the biodegradation effect.

E-015 내열성 폴리락틱산 수지의 사출성형

대표적인 생분해성 플라스틱 중 하나는 폴리락틱산(PLA)이다. 폴리락틱산은 식물에서 추출한 생분해성 플라스틱 수지이다. 다음 합성 과정을 통해 화학적으로 합성되고 생분해 효과를 이용하여 재활용된다.

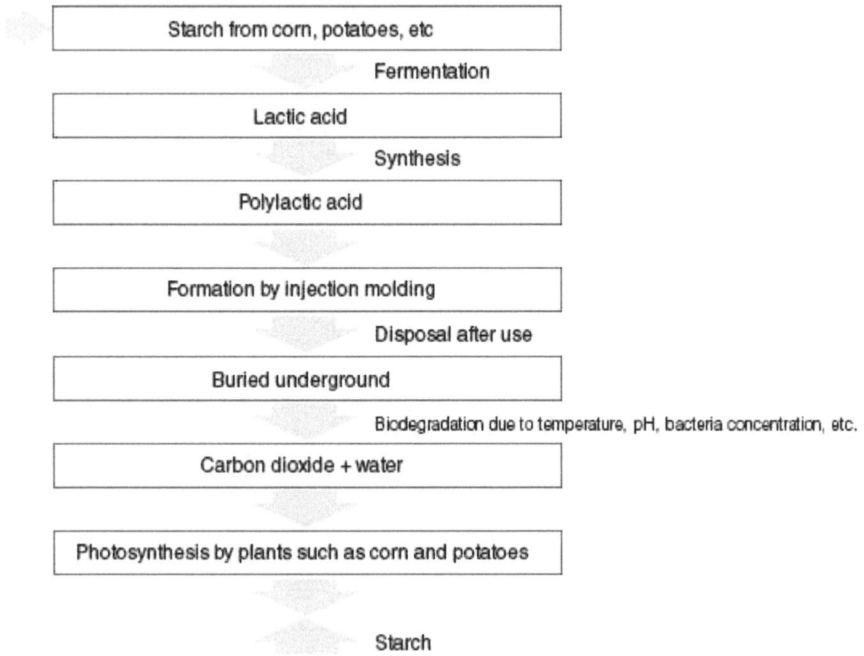

Polylactic acid is relatively hard and has about the same mechanical strength as polystyrene. Although it is being used in industrial applications for about ten years, since the strength and heat resistance are insufficient, it has not been used much in industrial products of high added value.

However, recently, polylactic acids having good heat resistance have been developed by the application of nano composites technology. Since such materials can withstand deflection temperatures under load of up to 120℃, it has become possible to clean them in hot water, or to use them continuously in an environment of steam.

Therefore, its application has been opened up to one-way food containers (disposable food containers), returnable food containers (food containers used repeatedly such as in school lunches, etc.), personal computer enclosures, office automation equipment enclosures, bottle containers, packaging containers, etc.

폴리락틱산은 비교적 단단하고 폴리스티렌과 거의 동일한 기계적 강도를 가지고 있다. 약 10년 동안 산업용으로 사용되고 있지만 강도와 내열성이 부족하기 때문에 고 부가 가치 산업 제품에는 그다지 많이 사용되지 않았다.

그러나 최근 나노 복합체 기술을 응용하여 내열성이 우수한 폴리락틱산이 개발되고 있다. 이러한 재료는 최대 120℃의 하중에서 처짐 온도를 견딜 수 있기 때문에 뜨거운 물로 세척하거나 연속적으로 증기 환경에서 사용할 수 있다.

따라서, 일회용 식품 용기(일회용 식품 용기), 회수 가능한 식품 용기(학교 점심에 반복적으로 사용되는 식품 용기 등), 개인용 컴퓨터 류, 사무 자동화기기 류, 병 용기, 포장 용기 등의 용도로 사용된다.

In order to obtain the heat resistance property of polylactic acid, it is necessary to maintain the mold cavity surface temperature at about 110 to 120℃. Therefore, a cartridge heating apparatus or an oil temperature control mechanism is mandatory in molds.

Further, heat resistant polylactic acid has a large shrinkage ratio (about 12 to 20/1000), and has the property of strongly clinging to the core. Mold design and machining, and surface treatment that make the mold releasing resistance small become necessary. Therefore, it is very much necessary to consider the ejection structure.

In particular, ejection is extremely difficult in the case of deep molded product such as cases, containers, etc. Various types of technologies are being developed regarding the mold structure and injection molding methods that make it possible to realize smooth ejection of molded items as well as their heat resistance.

폴리락틱산의 내열성을 얻기 위해서 금형 캐비티 표면 온도를 약 110 ~ 120℃로 유지할 필요가 있다. 따라서 금형에는 카트리지 가열 장치 또는 오일 온도 제어 장치가 필수적이다.

또한, 내열성 폴리락틱산은 수축률이 크고(약 12 ~ 20/1000), 강하게 코어에 달라붙는 성질을 갖는다. 이형 저항을 작게 만드는 금형설계 및 기계 가공 및 표면 처리가 필요하다. 따라서 취출구조를 고려하는 것이 매우 요구된다.

특히, 케이스, 용기 등의 깊이가 깊은 성형품의 경우, 취출이 극히 어렵다. 성형품의 내열성뿐만 아니라 원활한 배출을 가능하게 하는 금형 구조 및 사출성형 방법에 관해서는 여러 가지 기술이 개발되고 있다.

F. 금형부품소재

CHAPTER F

F 001 The Chemical Composition of Steel for Mold Manufacturing

The steel used in molds for plastic molding have ferrite-carbon alloy (which is normally called steel) as the basic material. It is helpful to know the chemical composition of some typical types of steel as a basic knowledge, because that will become useful when considering the heat treatment and mechanical characteristics, etc.

A table of that data is given below.

F-001 금형 제조용 철강의 화학 성분

플라스틱 성형용 금형에 사용되는 강재는 기본 재료로 페라이트-탄소 합금 (일반적으로 강재로 불림)이 있다. 열처리 및 기계적 특성 등을 고려할 때 유용하게 쓰일 수 있기 때문에 일부 기본적 강재의 화학적 조성을 기본적으로 이해하는 것이 도움이 될 것이다.

해당 데이터의 테이블은 아래와 같다.

Steel name	Symbol	Content ratio of chemical constituents (%)									
		C	Si	Mn	P	S	Ni	Cr	Mo	W	V
Rolled steel for general structures	[SS400] [1018 Steel Equivalent] [1.0040_ Ust.42.2]				0.06	0.06					
Carbon steel for mechanical structures	[S50C] [1049 Steel] [1.126_C50E (Ck50)]	0.45 -0.55	0.15 -0.40	0.40 -0.85	0.035	0.04					
Chrome molybdenum steel	SCM3	0.33 -0.38	0.15 -0.35	0.60 -0.85	0.03	0.03		0.90 -1.20	0.15 -0.35		
Stainless steel	SUS23	0.25 -0.40	0.75	1.00	0.04	0.03		12 -14			
Carbon tool steel	[SK5] [W1-8] [1.1525_ C80W1]	0.80 -0.90	0.35	0.50	0.03	0.03					
Alloy tool steel	[SKS3] [A1 or 01 Tool Steel] [1.2510_ 100MnCrW4]	0.90 -1.00	0.35	0.90 -1.20	0.03	0.03		0.50 -1.00		0.50 -1.00	

Steel name	Symbol	Content ratio of chemical constituents (%)									
		C	Si	Mn	P	S	Ni	Cr	Mo	W	V
Alloy tool steel hot die steel	[SKD61] [H13 Tool Steel] [1.2344_ X40CrMoV5 -1]	0.32 -0.42	0.80 -1.20	0.50	0.03	0.03		4.50 -5.00	1.00 -1.50		0.8 -1.2
Alloy tool steel cold die steel	[SKD11] [D2 Tool Steel] [1.2379_ X155CrVMo12 -1]	1.40 -1.60	0.40	0.50	0.03	0.03		11.0 -13.0	0.80 -1.20		0.2 -0.5

F 002 What is the Elastic Modulus of Steel?

"Elastic modulus" is a material property that indicates the strength or elasticity of the steel materials used for making mold parts. The elastic modulus is also called the "Young's modulus" usually. The elastic modulus is the coefficient of proportionality between the "strain" and the "tensile stress" when the steel material is pulled. This relationship can be expressed by the following equation.

$$\sigma = E \times \varepsilon$$

ε : Epsilon σ : Sigma

F-002 스틸의 탄성 계수는 무엇인가?

탄성 계수는 금형 부품을 제조하는 데 사용되는 강재의 강도 또는 탄성을 나타내는 재료 특성이다. 탄성 계수는 일반적으로 "영 계수(Young 's modulus)"라고도 한다. 탄성 계수는 강재를 잡아당길 때의 "변형률"과 "인장 응력"사이의 비례 계수이다. 이 관계는 다음 방정식으로 나타낼 수 있다.

이 관계는 다음 식으로 표현 될 수 있다.

$$\sigma = E \times \varepsilon$$

ε : 엡실론 σ : 시그마

	Unit
elastic modulus : E	kgf/cm² or Pa
Strain : ε	%
Tensile stress : σ	kgf/cm² or Pa

In other words, "stress is proportional to strain".

The physical value of the elastic modulus is determined by the type of the metallic material. In general, a material with a larger value for the elastic modulus has a higher tensile stress or rigidity.

The data of the elastic modulus is shown in Table 1 for some typical metallic materials.

즉, "응력은 변형률에 비례 한다".

탄성 계수의 실제 값은 금속 재료의 종류에 의해 결정된다.

탄성률의 물리적인 값은 금속 재료의 유형에 의해 결정된다. 일반적으로 탄성률에 대한 값이 더 큰 재료는 보다 높은 인장 응력 또는 강성을 갖는다.

일반적인 금속 재료에 대한 탄성 계수의 데이터가 표 1에 나와 있다.

	Elastic modulus E	
	(kgf/cm²)	**(MPa)**
Low carbon steel	210×10^4	20.59×10^4
S50C	210×10^4	20.59×10^4
Pre-hardened steel (SCM440 series)	203×10^4	19.9×10^4
SDK11	210×10^4	20.59×10^4
Brass	63×10^4	6.17×10^4
Copper	105×10^4	10.29×10^4
Aluminum	68×10^4	6.67×10^4
Super duralumin	73×10^4	7.16×10^4

F 003 Tempering of SKD61

Among the alloy tool steels, the so called hot work die steel (JIS name is SKD61) is being used as the material for the cavity or the core. It is also considered precious as a material for thin core pins because it has a relatively high hardness, withstands wear, and also has relatively high resistance to shock.

In order to bring out its excellent characteristics, SKD61 needs quenching, and after quenching, it is tempered to stabilize the metallographic texture and to improve its toughness. However, since it is known that depending on the conditions of tempering SKD61 causes change of dimensions and reduction of hardness, unforeseen failures can result if the tempering is done without understanding this trend. Therefore, in this course, the tempering characteristics of SKD61 are explained below.

1. Reduction in the hardness due to tempering

The hardness of SKD61 decreases in various cases by tempering after it is quenched. For example, if a material quenched at 1030℃ is tempered, the following changes occur depending on the tempering temperature. (The following data is an actual example, and there will be changes depending on the size of the work and the plate thickness.)

Therefore, since the hardness decreases suddenly if the tempering is done at an unnecessarily high temperature, care should be taken when resistance to wear is necessary.

F-003 SKD61의 템퍼링

합금 공구강 중 캐비티 또는 코어의 재료로서 열간 가공용 다이 스틸(JIS 이름 : SKD61)이 사용되고 있다. 그것은 또한 경도가 비교적 높고 마모를 견디며 충격에도 비교적 높은 내성을 갖기 때문에 얇은 코어 핀의 소재로 고려된다.

탁월한 특성을 이끌어 내기 위해 SKD61은 담금질이 필요하며, 담금질 후 금속 조직을 안정시키고 인성을 개선하기 위해 템퍼링 처리 한다. 그러나 템퍼링 조건에 따라 치수의 변화 및 경도의 감소를 일으키는 것으로 알려져 있으므로 이러한 경향을 이해하지 않고 템퍼링을 수행하면 예기치 않은 결함이 발생할 수 있다. 따라서 본 강좌에는 SKD61의 템퍼링 특성에 대해 설명한다.

1. 템퍼링에 의한 경도 저하

SKD61의 경도는 담금질 후 템퍼링에 의해 다양한 경우에 감소한다. 예를 들어 1030℃에서 담금질 된 재료가 템퍼링되면 템퍼링 온도에 따라 다음과 같은 변화가 발생한다.(다음 데이터는 실제 예이며, 작업 물의 크기와 판 두께에 따라 변화 될 수 있다.)

따라서 불필요한 고온에서 템퍼링을 하면 경도가 급격히 떨어지므로 마모 저항이 필요한 경우에는 주의를 기울여야한다.

2. Changes in the dimensions due to tempering

The external dimensions of the work change, when SKD61 is tempered after it is quenched. Depending on the tempering temperature, the external dimensions can become larger or smaller. If additional machining is not to be done after tempering, it is necessary to carry out the machining work before quenching taking into considerations the change in the dimensions after tempering.

2. 템퍼링에 의한 치수 변화

작업물의 외형 치수는 SKD61이 담금질 된 후 템퍼링 될 때 변경된다. 템퍼링 온도에 따라 외부 치수가 커지거나 작아 질 수 있다. 템퍼링 후에 추가 가공을 하지 않으려면 템퍼링 후 치수 변화를 고려하여 담금질 전에 가공 작업을 수행해야 한다.

F 004 Linear Expansion Coefficients of Materials

F-004 재료의 선팽창 계수

Although materials based on carbon steels are used for molds, some times non-ferrous metals of non-metallic materials are used for molds for the purpose of thermal insulation. At the time of assembling molds, although it is possible to adjust the dimensions at room temperature, since the temperature is increased and decreased during injection molding, the parts tend to undergo thermal expansion (either linear expansion or volume expansion). If the margin for expansion is not considered, the operation of the mold can become bad or the parts can break. The following data is available for the linear thermal expansion coefficients of materials.

탄소강을 기반으로 하는 재료가 금형에 사용 되더라도, 비금속 재료의 비철금속은 단열 목적으로 금형에 사용된다. 금형 조립 시 실온에서 치수를 조정할 수는 있지만 사출성형 중에 온도가 상승 및 하강하기 때문에 부품이 열팽창 (선 팽창 또는 부피 팽창)되는 경향이 있다. 팽창 여유가 고려되지 않은 경우 금형의 작동이 불량 해 지거나 부품이 파손될 수 있다. 다음 데이터는 재료의 신형 열팽창 계수를 나타낸 것이다.

Material name	Linear thermal expansion coefficient (Unit: $x10^{-6}/K$, 293K = 20℃)	Material name	Linear thermal expansion coefficient (Unit: $x10^{-6}/K$, 293K = 20℃)	Material name	Linear thermal expansion coefficient (Unit: $x10^{-6}/K$, 293K = 20℃)
S55C	11.7	Platinum	8.9	Carbon	3.1
SKD11	11.7	Tin	20	Tungsten	4.5
Pre-hardened steel	11.5	Silicon	2.6	Porcelain	6.8
18-8 stainless steel	17~18	Zirconia	5.4	Marble	3~15
Nickel steel	0.9	Diamond	1.0	Brick	3~6
Iron	11.8	Brass	18~23	Glass	9
Nichrome	18	Bronze	17.3	Quartz glass	0.5
Ultra Invar	−0.01	Titanium	8.2	Concrete	7~13
Duralumin	23	Silver	18.9	Acrylic resin	70~90
Copper	16.5	Gold	14.2	Bakelite	21~33

F 005 Deformation of Mold Components

After the molds are used actually for injection molding for some time, the mold components that use carbon steels and alloy tool steels and that have been heat-treated may generate warps, deformations, or their dimensions may have increased slightly. The phenomenon of changes occurring in the dimensions after the passage of time in this manner is called "aging".

F-005 금형 부품의 변형

금형이 실제로 사출성형에 사용된 후에는 탄소강 및 합금 공구강으로 만들어진 부품들은 휘어짐, 변형 또는 치수의 증가가 있을 수 있다. 이러 현상이 지속되어 치수가 변하는 현상을 "에이징"이라고 한다.

The main cause of changes with time occurring in carbon steels of alloy tool steels is known to be the expansion in volume caused when the austenite structures remaining during quenching changes to martensite structures.

The process of quenching is that of changing the austenite structure to the martensite structure by suddenly cooling from the quenching temperature (about 800℃, but this varies depending on the type of steel).

While the sudden cooling is done using water, oil, or a salt bath, although the conversion to martensite is promoted when the temperature is less than 0℃, if the temperature is not that low, a small quantity of austenite structure remains within the martensite structure. This remaining part is called "remained austenite".

It has been known that this remained austenite gradually changes to martensite structure with the passage of time as mentioned above, and at that time the volume also expands. As a consequence, if the quantity of remained austenite is large, it can be assumed that the trend is that of a large change with time.

In order to reduce the remained austenite, it is effective to carry out deep cooling treatment (subzero treatment) by creating a low temperature environment of about -80℃ using a coolant such as Freon (the user of Freon cannot be recommended due to environmental protection), and to carry out sudden cooling in that environment.

If liquid oxygen or liquid nitrogen is used, although it is possible to cool down to -180℃ to -190℃, since the cost becomes high this is only done under special cases.

합금 공구강 중에서 탄소강이 시간에 따라서 변화하는 주요 원인은 담금질 중에 잔류하는 오스테나이트 구조가 마르텐사이트 구조로 변화할 때 발생하는 부피의 팽창으로 알려져 있다.

담금질 과정은 담금질 온도에서 갑자기 냉각시켜 오스테나이트 조직을 마르텐사이트 조직으로 변화시키는 것이다. (약 800℃이지만 강철의 종류에 따라 다름).

물, 기름, 염욕을 통한 급속 냉각이 이루어지지만, 온도가 0℃ 미만이면 마르텐사이트로 전환되지만 온도가 낮지 않으면, 소량의 오스테나이트 조직이 마르텐사이트 구조 내에 남아있게 된다. 이 나머지 부분은 "잔류 오스테나이트"라고 한다.

이 잔류 오스테나이트는 전술 한 바와 같이 시간 경과에 따라 서서히 마르텐사이트 조직으로 변화하고, 그 시간에 부피가 팽창하는 것으로 알려저 있다. 결과직으로, 잔류 오스테나이트의 양이 크다면 그 경향은 시간에 따른 커다란 변화라고 추측 할 수 있다.

잔류 오스테나이트를 저감시키기 위해서 프레온 등의 냉매를 사용하여 약 -80℃의 저온 환경을 조성하여 깊은 냉각 처리(서브제로 처리)를 실시하는 것이 효과적이다.(프레온 사용자는 환경 보호) 그 환경에서 급속 냉각을 수행 할 수 있다.

액체 산소 또는 액체 질소를 사용하는 경우 -180℃ ~ -190℃까지 냉각 할 수는 있지만 비용이 많이 들기 때문에 특별한 경우에만 수행된다.

Therefore, in the case of components that should not undergo changes with time, it is recommended to carry out subzero treatment.

However, although the remained austenite would have become small in a subzero treated steel, at the same time, since even the hardness would have increased and the internal strain would have become large, appropriate annealing processing will be required. (There are recommended values for the conditions of annealing depending on the type of steel.)

따라서 시간 경과에 따른 변화가 없어야하는 부품의 경우, 서브제로 처리를 수행하는 것이 좋다.

그러나 잔류 오스테나이트는 서브제로 처리 강에서 작아졌지만 동시에 경도가 증가하고, 내부 변형이 커지므로 적절한 어닐링 처리가 필요할 것이다. (강의 유형에 따라 어닐링 조건에 대한 권장 값이 있다.)

F 006 Texture and Steel Materials

F-006 문양과 철강 재료

Texture is used for providing artificial undulation patterns or texturing on the surface of molded products.

"Texture" engraves undulations on the surface of the cavity similar to "etching", or "sandblasting".

In texture, etching, an acidic corrosive liquid is made to come into contact with and erodes the surface of the steel material under managed conditions.

When carrying out texturing, the technique is to pay attention to the following points.

1) The surface roughness of the cavity surface should be polished as finely as possible.

   Further, even the extent of polishing should be as uniform as possible.

Texture는 성형 제품 표면에 인공적인 패턴이나 무늬를 만들 때 사용된다.

"Texture"는 "식각" 또는 "샌드블라스팅"과 유사한 캐비티의 표면에 무늬를 새겨 넣는다.

무늬와 에칭에서 부식성 산성 액체는 일정 조건 하에서 강재의 표면과 접촉하여 침식시킨다.

무늬 제작 시에 다음과 같은 점을 주의해야한다.

1) 캐비티 면의 표면 거칠기는 가능한 미세하게 연마되어야 한다.

   또한, 연마의 정도도 가능한 균일해야한다.

2) The steel material should be of the same steel type as far as possible, and should be selected from the same lot.

If the steel type is different, the state of erosion may change in a subtle way. Even if the same erosion processing is done for SKD11 and the pre-hardened steel S55C, there may be subtle differences in the state of the undulations.

In addition, even if the same steel type is used, there is the danger of the finished state after texturing being different due to differences in the lot of the material or the direction of rolling of the material.

If the state of texturing becomes different, in order to carry out texturing again, it becomes necessary to carry out polishing again of the cavity surface which can lead to a huge loss of time.

2) 강재는 가능한 동일한 강재이어야 하며, 동일한 롯트에서 선택되어야한다.

강재 유형이 다르면 침식 상태가 미묘하게 바뀔 수 있다. SKD11과 프리 하든강 S55C에 대해 동일한 침식 처리가 수행 되더라도 무늬 상태에 미묘한 차이가 있을 수 있다.

또한, 동일한 강재를 사용해도 재료의 롯트나 압연 방향의 차이에 의해 무늬 가공 후의 마무리 상태에 차이가 있을 수 있다.

무늬의 상태가 다르게 되면 다시 무늬 작업을 수행하기 위해 시간이 엄청나게 길어질 수 있고, 캐비티 표면의 연마를 다시 수행해야한다.

3) Particularly careful polishing is necessary for surfaces prepared by electric discharge machining.

Surfaces that are prepared by electric discharge machining will have a machining hardened layer that may have appeared, and hence it is recommended to carry out thorough polishing of such surfaces.

3) 방전 가공에 의해 준비된 표면에는 특히 주의해 깊은 연마를 해야 한다.

방전 가공 된 표면에는 가공 경화 층이 생길 수 있으므로 이러한 표면을 철저히 연마하는 것이 좋다.

F 007 Thermal Conductivity(λ) of Materials

F-007 재료의 열전도

In molds or plastic materials, the physical phenomenon of processing or deforming is done by putting in or taking out heat by heating or cooling. The property of conveying heat is determined by the structure of the material. Thermal conductivity is one of the properties as being easy to convey (or hard to convey) heat through the material.

Thermal conductivity is an index which expresses how much quantity of heat is conveyed within a unit time, over a unit length, and with a unit temperature change. The most commonly used symbol for this is λ (lambda). The unit is kcal/m • h • ℃.

금형 또는 플라스틱 재료에서 가공 또는 변형의 물리적 현상은 열의 가열 또는 냉각에 의해 수행된다. 열전달은 재료의 구조에 의해 결정된다. 열전도성은 소재를 통해 열을 쉽게 전달할 수 있는 (또는 전달하기 어렵다) 특성 중 하나이다.

열전도성은 단위 시간 내에 단위 길이에 걸쳐 단위 온도 변화와 함께 얼마나 많은 양의 열이 전달되는지를 나타내는 지표이다. 이를 위해 가장 일반적으로 사용되는 기호는 λ (람다)이다. 단위는 kcal/m • h • ℃ 이다.

Material name	λ	Material name	λ	Material name	λ	Material name	λ
Aluminum	175	Magnesium	147	Carbon steel	31-46	Gold	267
Duralumin	141	Lead	30	Nickel steel	12-62	Silver	360
Iron	58	Zinc	96	Invar	9.2	Platinum	60
Cast iron	41	Tin	55	Chrome steel	27-62	Cotton	0.051
Copper	332	Glass	0.65	Brass	52	Concrete	0.7-1.2
Bronze	22	Mica	0.7-1.2				

F 008 Nickel

F-008 니켈

Nickel can be used in materials of special metals or for electroplating on metal surfaces. Nickel is a silvery shiny metal.

니켈은 특수 금속 재료 또는 금속 표면의 전기 도금에 사용할 수 있다. 니켈은 은빛이 나는 금속이다.

Element symbol	Ni
Element number	28
Specific gravity	8,902
Melting point	1,453℃
Specific heat	0.105cal/gf · ℃
Linear coefficient of expansion	7.31×10^{-6}
Thermal conductivity	0.125cal/cm · sec · ℃
Crystal lattice	Face centered cubic structure

The purpose of use of nickel in molds is mainly to exploit its resistance to corrosion. Nickel is extremely stable in air, and does not react to oxidation or retardation. It is also stable in water or brine and does not generate rust. In particular, it exhibits remarkable resistance to sulfuric acid and hydrochloric acid

In addition, nickel is also used in a condition in which it is alloyed with other metallic elements.

금형에 니켈을 사용하는 목적은 주로 부식에 대한 저항력을 이용하기 위해서다. 니켈은 공기 중에서 극도로 안정적이며, 산화 또는 방해물에 반응하지 않는다. 그것은 또한 물 또는 소금물에 안정하고 녹이 발생하지 않는다. 특히, 황산 및 염산에 대하여 우수한 내성을 나타낸다.

또한 니켈은 다른 금속 원소와 합금된 조건에서도 사용된다.

Ni-Cu	Used as the material for thermocouples
Ni-Al	-
Ni-Cr	Used as Nichrome wire of heaters
Ni-Fe alloy	Used as the corrosion resistant alloy Hastelloy.
Ni-Cr-Fe alloy	Used as the heat resistant and corrosion resistant alloy Inconel.

As a day-to-day application, the corrosion resistance of nickel is utilized in coins. In molds, nickel is used very often as films on the surfaces of molds in order to prevent gases or corrosive constituents being generated from molten plastics, cavity generation by adhered moisture, or to prevent corrosion of the core by oxidization.

In carbon steel, it is used as an alloy along with chromium (Cr), and is used widely as stainless steel for cavities, cores, and ejector pins, etc.

Since nickel is costlier than carbon steel, using it in the form of electroplated films or alloys can be said to be a rational method of use.

일상적으로 사용되는 니켈의 내부식성은 동전에 사용된다. 금형에서 니켈은 용융수지, 캐비티의 수분 등으로 발생하는 가스 또는 부식성 성분을 막거나 방지하기 위한 피막으로 매우 자주 사용된다.

탄소강에는 크롬(Cr)과 함께 합금으로 사용되며, 캐비티, 코어 및 이젝터 핀 등을 위해 스테인리스 강으로 널리 사용된다.

니켈은 탄소강보다 값이 비싸기 때문에 전기 도금 막이나 합금 형태로 사용하는 것이 합리적인 사용 방법이라고 할 수 있다.

F 009 Maraging Steel

F-009 마레이징 강

Maraging steel is one type of special steel used for the for plastic injection mold.
Usually, when quenching and heat treating special steels, although the solidification is done by sudden cooling from a temperature higher than the A1 transformation point (723℃) after obtaining the martensite structure, considerably large residual stresses remain in the material due to sudden cooling, and also there will be dimensional deformations. The maraging a type of steel in which the steel is hardened without taking such risks.

The following typical types are present maraging steels.

마레이징 강은 플라스틱 사출 금형에 사용되는 특수강의 한 유형이다.

보통 특수강을 담금질 및 열처리 할 때 A1을 얻은 후 A1 변형 점(723℃)보다 높은 온도에서 급냉시켜 냉각을 하지만 마르텐사이트 구조는 갑작스런 냉각으로 인해 재료에 상당한 잔류 응력이 남아 있고 치수 변형도 있을 수 있다. 마레이징 강은 그러한 단점을 고려하지 않아도 되는 강철이다.

다음과 같은 전형적인 유형의 마레이징 강재가 있다.

1) SUS630 type
2) Maraging steel
3) Non-magnetic steel

1) SUS630 타입
2) 마레이징 강
3) 비자성 강

The principles by which the maraging steel becomes hard is that, the steel is supplied in the soft state after processing in a method called solution heat treatment. Next, after the material is machined into the component shapes such as the cavity or the core, etc., it is heated and left to cool. When this is done, a phenomenon called age hardening occurs, and the hardness of the steel increases naturally. In the age hardening process, compositions called intermetallic compounds are precipitated, and this increases the hardness.

Since heat treatment is simpler than quenched and annealed steel, it is also easy to stabilize the dimensions, and maraging steel is a useful item depending on the application.

The appropriate temperature of age hardening depends on the type of steel, and the optimum processing is selected following the respective guidelines.

The features of the different steel materials are given below.

마레이징 강이 견고해지는 원리는 우선, 강제가 용체화 처리라 불리는 방법으로 가공 된 후 부드러운 상태로 공급되기 때문이다. 다음에 캐비티나 코어 등의 부품 형상으로 가공 한 후, 가열하여 냉각시킨다. 이것이 끝나면 시효 경화라는 현상이 일어나고 강재의 경도는 자연스럽게 증가한다. 시효 경화 공정에서 금속 간 화합물이라 불리는 조성물이 침전되고, 이것은 경도를 증가시킨다.

열처리가 풀림 및 어닐링 된 강보다 간단하기 때문에 치수를 안정화하기 쉽다. 마레이징 강은 사용 용도에 따라 유요한 아이템이다.

시효 경화의 적정 온도는 강재의 종류에 따라 다르며 각각의 지침에 따라 최적 가공이 선택된다.

다른 강재의 특징은 다음과 같다.

1. SUS630 type

This steel is obtained by modifying the stainless steel SUS 630. A steel material with a hardness in the range of 35 to 40 HRC is supplied, and very often it would already have been age hardened. Its resistance to corrosion is high and is suitable for the die steels of plastics having high corroding property.

1. SUS630 형

이 강은 스테인리스 강 SUS630은 개질하여 얻어진다. 35 ~ 40 HRC의 경도를 가진 강재가 공급되며, 이미 오래 전부터 경화되어 있다. 부식에 대한 내성이 높고 내식성이 우수한 플라스틱의 다이 스틸에 적합하다.

2. Maraging steel

The specularity of this steel is very good. This is used very frequently for the cavities of optical components, or lens cavities. This material has both high hardness and high strength, and also has toughness. This steel is used also for thin core pins and long and thin components.

2. 마레이징 강

이 강의 반사율은 아주 양호하다. 이것은 광학 구성 요소 또는 렌즈 캐비티에 매우 자주 사용된다. 이 재료는 높은 경도와 높은 강도를 모두 가지고 있으며 인성도 양호하다. 이 강재는 얇은 코어 핀과 길고 가는 부품에도 사용된다.

3. Non-magnetic steel

This is a steel does not stick to a magnet and is used for plastic magnets. Since a magnet has a high hardness and wears out the steel, a high hardness is required in the steel. By carrying out solution hardening, the hardness increases and it is possible to obtain a steel having non-magnetic properties.

3. 비자성 강

이 강은 자석에 달라붙지 않으며 플라스틱 자석에 사용된다. 자석은 경도가 높고 강철을 마모시키기 때문에 강재에는 높은 경도가 요구된다. 용제 경화에 의해 경도가 상승하여 비자성을 갖는 강을 얻을 수 있다.

G. 금형부품설계

CHAPTER G

G 001 Cost Analysis of Plastic Molded Article

In a company manufacturing and selling plastic molded articles, it is necessary to carry out a careful analysis of what expenses are actually being incurred where and for what item and to be of use to the management. This kind of analysis is called cost analysis. If a cost analysis is made, it can be understood that the following items will have to be accounted as the production expenses of the molded articles.

Cost analysis of plastic molded items:

- Material expenses
- Direct overhead (workers)
- Indirect overhead (office workers, managers)
- Insurance
- Mold depreciation
- Machine depreciation (molding machine)
- Machine depreciation (peripheral equipment)
- Maintenance expenses
- Building depreciation
- Electricity expenses
- Water expenses
- Defect countermeasure expenses
- Inspection expenses
- Packaging expenses
- Freight expenses

G-001 플라스틱 성형품의 원가 분석

플라스틱 성형품을 제조 및 판매하는 회사의 경우, 실제로 어떤 비용이 발생했는지, 관리 대상 품목 및 사용 대상을 신중히 분석해야한다. 이러한 종류의 분석을 비용 분석이라고 한다. 비용 분석이 이루어지면 성형품의 생산 비용으로 다음 항목을 고려해야 한다.

플라스틱 성형품의 원기 분석 :

- 자재 비용
- 직접 간접비 (근로자)
- 간접비 (회사원, 관리자)
- 보험료
- 금형 감가 상각
- 기계 감가 상각 (성형 장비)
- 기계 감가 상각 (주변 장비)
- 유지 보수 비용
- 빌딩 감가 상각비
- 전기 비용
- 수질 관리 비용
- 검사 비용
- 패키지 비용
- 운임 비용

In addition, the following expenses are accounted in some cases.

- Secondary processing expenses
- Assembly expenses
- Technology license fees (patent utilization royalties, etc.)

Further, the sales price will be the profit added to these expenses.
The ratio (%) of profits to the sales price is called the gross profit, and the ratio obtained after deducting taxes from the gross profit is called the net profit ratio (%). Being able to produce a large number of high profit ratio molded items is the path to good management.

또한 경우에 따라 다음 경비가 고려된다.

- 2차 처리 비용
- 조립 비용
- 기술 라이센스 비용(특허 사용료 등)

판매 가격은 이 비용에 더해진 이익이다. 판매 가격에 대한 이익 비율(%)을 총 이익이라고 하고, 총 이익에서 세금을 공제 한 후 얻은 비율을 순이익률(%)이라고 한다. 고수익률의 성형품을 대량으로 생산할 수 있다는 것은 훌륭한 관리 방법이다.

G 002 Initial Cost Structure of Molds

G-002 금형의 초기 비용 구조

The value of a mold is determined by the three factors of "Quality", "Delivery", and "Cost".

Among these, the "Cost" is constituted from initial cost and running cost.

In addition, it is very common that the initial cost is constituted by the following cost constituents.

금형의 가치는 "품질", "납품" 및 "원가"의 3 가지 요소에 의해 결정된다.

이 중 "비용"은 초기 비용과 운영 비용으로 구성된다.

또한 초기 비용은 다음과 같은 원가 구성 요소로 구성되는 것이 일반적이다.

(1) Raw material cost

The cost of purchasing the steel raw material for the mold.

(1) 원 재료비

금형을 위한 강재의 구매 비용

(2) Cost of purchasing standard components

The cost of purchasing standard components such as the ejector pin, sprue bush, etc.

(2) 표준 구성 요소 구매 비용

이젝터 핀, 스프루 부시 등과 같은 표준 부품을 구입하는 데 드는 비용

(3) Cost of purchasing the mold base

The cost of purchasing the mold base, and accessory components, additional machining, etc.

(3) 몰드베이스 구매 비용

몰드베이스 및 액세서리 부품, 추가 가공 등의 구입비용

(4) Mold design cost

The expenses incurred for designing the mold.

(4) 금형설계 비용

금형설계 비용

(5) Program preparation expenses

The expenses incurred for preparing the NC program.

(5) 프로그램 준비 비용

NC 프로그램 준비 비용

(6) Machining expenses

The expenses incurred for machining operations such as cutting, grinding, electric discharge machining, etc.

(6) 가공 비용

절삭, 연삭, 방전 가공 등과 같은 가공 작업에 소요되는 비용

(7) Surface treatment expenses

The expenses incurred for electroplating, PVD coating, etching, etc.

(7) 표면 처리 비용

전기 도금, PVD 코팅, 에칭 등에 소요되는 비용

(8) Finishing and assembly expenses

The expenses incurred for polishing, finishing, and assembly.

(8) 다듬질 및 조립 비용

연마, 다듬질 및 조립에 소요되는 비용

(9) Trial operation expenses

The expenses related to trial injection molding such as the cost of plastic materials, etc.

(9) 시운전 비용

플라스틱 재료비처럼 시험사출에 관련 된 비용

(10) Measurement expenses

The expenses incurred for measuring the dimensions of the mold components or the molded product.

(10) 측정 비용

금형 부품 또는 성형품의 치수를 측정 할 때 소요되는 비용

(11) Packaging and transportation cost

The expenses incurred for packaging and transporting the molds.

(11) 포장 및 운송 비용

금형을 포장하고 운반하는 데 드는 비용

(12) Trouble shooting expenses

The expenses incurred in trouble shooting such as design mistakes, machining mistakes, etc.

(12) 문제 해결 비용

설계 오류, 가공 오류와 같은 시험사출과 관련된 비용

(13) Instruction manual preparation cost

The cost of preparing the instruction manual for the mold and of preparing associated documents.

(13) 사용 설명서 준비 비용

금형에 대한 사용 설명서 준비 및 관련 문서 준비 비용

(14) Net profit

The net profit obtained by the company manufacturing the molds.

(14) 당기 순이익

회사가 금형을 제조하여 얻은 순이익

(15) Taxes

Corporate taxes, sales taxes, etc.

(15) 세금

법인세, 판매세 등

G 003 Design Concepts of Mold Design

The mold design is an intellectual work of realizing the idea of the mold so that the molded product is within the desired specifications and the cost is within the requested limit, and the items to be investigated can be classified into the following factors.

Functional design

- Maintaining the dimensions of the molded product
- Investigations of the runner and gate structure
- Investigations of the ejection method

Shape design

- Three dimensional shape design
- External appearance quality investigations
- Investigation of shape transfer, etc.

Strength design

- Cavity and core deformation, breakage
- Deformation of the mold base, etc.

Thermal design

- Investigation of the thermal shrinkage with respect to the mold
- Investigation of the cooling capacity of the molded product
- Investigation of the temperature controlling means

G-003 금형설계의 개념

금형설계는 금형 아이디어를 실현하는 성형품이 원하는 규격 내에 있고, 비용이 요구 한도 내에 있도록 하는 지적 작업이다. 조사 되는 설계 요소는 아래와 같이 분류 할 수 있다.

기능적 디자인

- 성형품의 치수 유지
- 런너 및 게이트 구조 조사
- 취출 방법 조사

형상 디자인

- 3차원 형상 설계
- 외관 품질 조사
- 형상 변경 등의 조사

강도 설계

- 캐비티 및 코어 변형, 파손
- 몰드베이스 변형 등

열관련 설계

- 금형에 대한 열 수축률 조사
- 성형품 냉각 능력 조사
- 온도 조절 수단 조사

Safety design

- Investigation of the hook bolt strength
- Investigation of the mold opening prevention structure, etc.

안전 설계

- 후크 볼트 강도 조사
- 형 개폐 방지 구조 등의 조사

Maintenance design

- Investigation of the maintenance parts structure
- Investigation of spare parts
- Investigation of ease of disassembling and assembling
- Investigation of repeatability

유지 보수 설계

- 유지 보수 부품 구조 조사
- 예비 부품 조사
- 해체 및 조립의 용이성에 대한 조사
- 반복성 조사

Cost design

- Investigation of the mold cost
- Investigation of the molded product cost
- Investigation of using standard parts, etc.

비용 설계

- 금형 비용 조사
- 성형품 가격 조사
- 표준 부품 등의 조사

Mold design is not merely creating a space for producing the shape of the molded product, but it is very important to carry out various studies from the above viewpoints.

금형설계는 단순히 성형품의 형상을 만드는 공간을 만드는 것이 아니며, 상기 관점에서 여러 가지 연구를 하는 것이 중요하다.

G 004 Merits or Demerits of Standardization for Mold Design and Fabrication

In the world of design and fabrication of molds for plastic injection molding, it can be said that standardization spread relatively faster than other fields of machine design and fabrication.

Misumi has had a role to play in this matter, and we would like to discuss here again the merits and demerits of standardization of design and fabrication by making quantitative comparisons.

1. Merits of Mold Design and Fabrication

1) The design time can be made short

Because of this, the design delivery time becomes short, and even the design cost gets reduced.

2) It is possible to shorten the machining time, the finishing and adjustment times.

Because of this, the component preparation time gets shortened, and even the production cost gets reduced.

3) It is not necessary to increase the number of machines and equipment

Since it is possible to depend on purchasing, it is not necessary to increase the number of machines and equipment within the company. Therefore, the fixed expenses (cost of depreciation) need not be increased.

G-004 금형설계 및 제조 표준화의 장단점

플라스틱 사출금형설계 및 제조에서 표준화는 다른 기계 설계 및 제조 분야보다 상대적으로 빠르게 확산된다고 할 수 있다.

미스미(Misumi)는 이 문제에서 중요한 역할을 담당했으며, 양적 비교를 통해 설계 및 제조 표준화의 장단점을 다시 논의하고자한다.

1. 금형설계 및 제작의 장점

1) 설계시간을 단축 할 수 있다.

이 때문에 설계 전달 시간이 단축되고, 설계 비용 조차 줄어들게 된다.

2) 가공시간 단축, 다듬실 빛 조성시간 단축이 가능하다.

이 때문에 부품 준비시간이 단축되고, 생산 원가도 절감된다.

3) 기계 및 장비 수를 늘릴 필요가 없다.

구매에 의존 할 수 있기 때문에 회사 내의 기계 및 장비 수를 늘릴 필요가 없다. 따라서 고정비용(감가 상각비)을 늘릴 필요가 없다.

4) The interchangeability during maintenance gets enhanced

By using standard components, since it is possible to unify the specifications of replacement parts, the interchangeability becomes excellent.

4) 유지 보수시의 호환성이 향상된다.

표준 부품을 사용하여 교체 부품의 사양을 통일 할 수 있으므로 상호 호환성이 탁월하다.

5) Maintenance becomes easy in the case of exports to overseas destinations

At the time of maintaining molds exported to overseas destinations, if the same standard component can be procured locally, it becomes possible to carry out maintenance speedily.

5) 해외 수출의 경우 유지 보수가 용이하다.

금형을 해외로 수출 할 때 동일한 표준 부품을 현지에서 조달 할 수 있다면 신속하게 유지 보수를 수행 할 수 있다.

6) The cost of order management can be reduced.

The additional work load of ordering and associated accounting is eliminated, and as a result it is possible to reduce the cost.

6) 주문 관리 비용을 줄일 수 있다.

주문 및 관련 회계의 추가 작업 부담이 제거되고, 결과적으로 비용을 절감 할 수 있다.

2. Demerits of Mold Design and Fabrication

2. 금형설계 및 제작의 단점

1) It is likely that it becomes difficult to pass on the basic knowledge or mold design to the freshly recruited designers.

1) 새로 입사한 설계자에게 기본 지식이나 금형설계를 전수하는 것이 어려워 질 가능성이 있다.

2) The basic component production technology within the company may be lost.

2) 회사 내의 기본 구성 요소 생산 기술이 하락 될 수 있다.

In order to counterbalance the demerits of standardization, there are some companies that carry out design seminars within the company or internal production within the company of parts for repairs.

Considering the advantages and disadvantages and the losses and gains, it can be said that further progress in the standardization of design and fabrication of molds is definitely advantageous.

표준화의 단점을 보완하기 위해 회사 내에서 설계 세미나를 수행하거나 수리를 위해 부품을 회사 내에서 생산하는 일부 회사가 있다.

장점과 단점, 손실과 이득을 감안할 때 금형의 설계와 제조의 표준화가 더욱 진전되었다고 할 수 있다.

G 005 Strength Theory of Mold Design

While molds are precision machines for molding plastic, it is necessary to acquire appropriate strength in order to maintain its functions. In addition, since a mold becomes a heavy load when carrying out the operations of installing it or removing it, sufficient strength has to be acquired so that accidents causing injury to humans is not caused due to sudden problems such as the mold getting deformed or damaged due to its own weight while it is being carried.

In order to do this, it is not sufficient to ensure the strength based merely on experience or intuition, but it is important to carry out mold design using strength calculations based on the mechanics of materials.

In the design of plastic injection mold, it is necessary to utilize the data of the following experiments based on the mechanics of materials.

■ Static strength tests

- Tension test
- Bending test
- Shear test
- Compression test
- Buckling test
- Torsion test

■ Dynamic strength tests

- Impact test (Charpy impact test, Izod impact test)
- Fatigue test

■ Industrial tests

- Creep test
- Wear test, etc.

G-005 금형설계의 강도 이론

금형은 플라스틱 성형을 위한 정밀 기계이지만, 기능을 유지하려면 적절한 강도를 가져야 한다. 또한, 금형의 설치 작업을 수행하거나 제거할 때 중량 부하가 작용하기 때문에 금형이 변형되거나 손상되는 것과 같은 갑작스러운 문제로 인해 인체에 상해를 주는 사고가 야기되지 않도록 충분한 강도를 가져야 한다.

이를 위해서 단순히 경험이나 직감에 기초한 강도를 확보하는 것만으로는 충분하지 않지만 재료 역학에 기반한 강도 계산을 사용하여 금형설계를 수행하는 것이 중요하다.

플라스틱 사출 금형의 설계에는 재료 역학에 기초하여 다음 실험의 데이터를 활용할 필요가 있다.

■ 정적 강도 테스트

- 인장 시험
- 굽힘 시험
- 전단 시험
- 압축 시험
- 좌굴 시험
- 비틀림 시험

■ 동적 강도 시험

- 충격 시험 (샤르피 충격 시험 아이조드 충격 시험)
- 피로 시험

■ 산업용 테스트

- 크리프 시험
- 마모 시험 등

Empirical formulae that have been technically established should be used for the calculations related to the strength of materials, and it is necessary to use a safety margin factor regarding the results of calculations.

Further, although it goes without saying, the results of calculations should strictly follow the "ethics of engineers", and it is necessary to take decisions based on good intents.

Since the prevention of mold damage accidents and acquisition of safety depend upon the judgment by the designing engineer, one should always be strict with oneself so that too much weight is not given to designs for cost reduction thereby blunting one's own most important decisions. This is something a mold design engineer should always remember.

기술적으로 확립된 경험식은 재료 강도와 관련된 계산에 사용되어야하며, 계산 결과와 관련하여 안전율을 사용해야한다.

또한, 계산 결과는 "엔지니어의 윤리"를 엄격히 준수해야하며, 좋은 의도를 바탕으로 의사 결정을 내릴 필요가 있다.

금형 사고 예방 및 안전 확보는 설계 엔지니어의 판단에 의존하기 때문에 비용 절감을 위해 디자인에 너무 많은 무게가 주어지지 않도록 항상 자신에게 엄격해야하고, 따라서 이러한 사항은 금형설계 엔지니어가 항상 기억해야 한다.

G 006 How is the Rigidity of a Mold Increased?

G-006 금형의 강도는 어떻게 증가하나?

A plastic injection mold is subjected to high internal pressure at the time of filling the molten plastic, and also, it is subjected to high compression stress at the time of clamping the mold. In addition, if the mold becomes large, it can also be subjected to a bending stress due to its own weight. In order to make sure that the mold does not get deformed or broken due to the external stresses or stresses due to its own weight, it is necessary to strengthen the rigidity of the mold.

Here, let us understand from the basics what rigidity is.

플라스틱 금형은 용융수지의 충전 시에 고압의 내부 압력이 가해지고, 또한, 금형 클램핑 시에 높은 압축 응력을 받게 된다. 금형이 커지면 자중에 의해 굽힘 응력을 받을 수도 있다. 금형이 자중에 의한 외부 응력이나 응력으로 인해 변형되거나 파손되지 않도록 하기 위해서 금형의 강성을 강화할 필요가 있다.

여기서, 우리가 강성이 무엇인지 기초부터 이해하자.

Rigidity is the resistance to deformation when subjected to a load. The modulus of longitudinal elasticity E and the modulus of transverse elasticity G of the material affect the rigidity. A material for which the value of E or G is large can be said to have a high rigidity. In other words, it exhibits strong resistance to bending or twisting. In more easy to understand terms, the material is difficult to bend, and also has a very small deflection.

For example, while the value of E for SCM440 series pre-hardened steel is 203×10^4 (kgf/cm^2) but the value of E for SKD11 (cold rolled die steel) is 210×10^4 (kgf/cm^2), it can be said that SKD11 is more rigid.

Explaining in more detailed terms, rigidity can be "bending rigidity" or "twisting rigidity". "Bending rigidity" is particularly more important in the case of the molds for plastic injection molding.

Bending rigidity (flexural rigidity) indicates the resistance to bending due to a bending load. In general, the bending rigidity is expressed by "E , I". (I is moment of inertia of area. For details, see the previous course.) In order to increase the bending rigidity, it is necessary to make the product E , I large. In other words, selecting a material with a large value of the modulus of longitudinal elasticity E and also adopting a cross-sectional shape that makes the moment of inertia of area I large results in making the bending rigidity high.

If the structure has a high bending rigidity, even the deflection becomes small and it is also possible to resist breakage due to bending deformations.

강성은 하중을 받는 변형에 대한 저항이다. 종탄성계수 E와 재료의 횡탄성계수 G는 강성에 영향을 미친다. E 또는 G의 값이 큰 재료는 높은 강성이라고 말할 수 있다. 즉, 굴곡이나 뒤틀림에 대한 강한 저항력을 나타낸다. 보다 이해하기 쉽게 말하면 소재는 구부리기 어렵고, 또한 매우 작은 처짐을 가지고 있다.

예를 들어, SCM440 시리즈의 프리하든 강은 E 값이 203×10^4 (kgf/cm^2)이지만 SKD 11(냉간 압연 강재)의 E 값은 210×10^4(kgf/cm^2)일 수 있다. SKD11은 더욱 강하다 말할 수 있다.

더 자세히 설명하면 강도는 "굽힘 강도"또는 "비틀림 강도"가 될 수 있다. "굽힘 강도"는 플라스틱 사출 금형의 경우 특히 더 중요하다.

굽힘 강도(굽곡 강도)는 굽힘 하중에 의한 굽힘 저항을 나타낸다. 일반적으로, 굽힘 강도는 "E, I"로 표시된디.(I는 단면 2차 모멘드로 이전 과정을 참조) 굽힘 강도를 높이기 위해서 제품 E나 I 값을 크게 할 필요가 있다. 즉, 종탄성계수(E)의 값이 큰 재료를 선택하는 동시에 단면 (I)의 단면 2차 모멘트를 크게 하는 단면형상을 선택함으로써 굽힘 강성을 높게 할 수 있다.

구조체가 높은 굽힘 강성을 가지면 휨이 작아지거나 굽힘 변형으로 인한 파손에도 견딜 수 있다.

G 007 Mold Base

Mold base is a comprehensive name used for the parts for containing the cavity for plastic injection mold, and also has the role of directly installing the mold to the plastic injection molding machine.

Mold base is a set of parts that constitute the outer periphery part of a plastic injection mold, and is constituted mainly from the following parts.

1) Fixed half retainer plate
2) Fixed half mold plate
3) Moving half mold plate
4) Spacer blocks
5) Ejector plate (top)
6) Ejector plate (bottom)
7) Moving half retainer plate
8) Runner stripper plates (in the case of a 3-plate structure)

Although previously the constituent parts of a mold base were all designed and manufactured as required, standard mold bases have recently come into wider use and are being used all over the world. In the case of large sized molds or small sized molds, even at present, they are being designed individually for each mold.

Although the standards for mold bases have been prepared in Japan in metric units, they are still being prepared in inch units in the U.S.A. In Europe, as in Japan, it is common to prepare them in metric units.

G-007 몰드베이스

몰드베이스는 플라스틱 사출성형용 캐비티를 포함하는 부품에 사용되는 포괄적인 이름이며, 플라스틱 사출성형기에 직접 설치된다.

몰드베이스는 플라스틱 사출금형의 외부를 구성하는 부품 세트이며, 주로 다음의 부품으로 구성된다.

1) 고정측 설치판
2) 고정측 형판
3) 이동측 형판
4) 스페이서 블록
5) 이젝터플레이트 (상)
6) 이젝터 플레이트 (하)
7) 가동측 설치판
8) 런너 스트리퍼 플레이트 (3매 구조의 경우)

예전에는 몰드베이스의 구성 부품이 모두 필요에 따라 설계되고 제조되었지만 표준 몰드베이스가 최근에 널리 사용되어 전 세계에서 사용되고 있다. 큰 크기의 금형 또는 작은 크기의 금형의 경우에는 현재까지도 금형마다 개별적으로 설계되고 있다.

몰드 베이스에 대한 기준은 미터법 단위로 일본에서 준비되었지만 미국에는 여전히 인치 단위로 준비되고 있다. 유럽과 마찬가지로 일본에서도 미터 단위로 준비하는 것이 일반적이다.

The following two types of structures are the most co mmonly used ones for the structure of a mold base.

다음 두 가지 유형의 구조는 몰드베이스 구조에 가장 일반적으로 사용되는 구조이다.

1) 2-plates structure
2) 3-plates structure

1) 2매 플레이트 구조
2) 3매 플레이트 구조

The selection between these two structures is determined by the method of the gate used. When adopting a pin point gate structure, always the structure 2) is used. In the case of a side gate or a tunnel gate, the structure 1) is used normally.

이 두 구조 사이의 선택은 사용된 게이트의 방법에 의해 결정된다. 핀 포인트 게이트 구조를 채택 할 때 항상 구조 2)가 사용된다. 사이드 게이트 또는 터널 게이트의 경우에는 구조 1)이 주로 사용된다.

The material for the constituent parts of the mold base is generally the carbon steel for machine construction (S55C, 220C, etc.,) and is used most often in the non-hardened condition. In special applications, pre-hardened steel, or stainless steel, or an aluminum alloy is used some times. A mold based uses in combination with accessory parts like guide pins, guide bushes, return pins, etc.

몰드베이스 재료는 일반적으로 기계 구성 용 탄소강 (S55C, 220C 등)이며, 경화되지 않은 상태에서 가장 많이 사용된다. 특수 용도에는 사전 경화 된 강철 또는 스테인레스 스틸 또는 알루미늄 합금이 가끔 사용된다. 몰드 기반은 가이드 핀, 가이드 부시, 리턴 핀 등과 같은 부속 부품과 함께 사용된다.

G 008 Moment of Inertia of area Mold Components

G-008 금형 부품의 단면 2차 모멘트

The moment of Inertia of area mold components which is important for forecasting bending is explained below.

굽힘 예측에 중요한 금형 부품의 단면 2차모멘트는 아래에 설명되어 있다.

At the time of carrying out the strength calculations of a mold for plastic injection molding, the term "moment of inertia of area" appears very frequently. Let us understand again what "moment of inertia of area" is

플라스틱 사출금형의 강도 계산을 수행할 때, "단면 2차 모멘트" 라는 용어가 매우 자주 나타난다. "단면 2차 모멘트"가 무엇인지 다

so that it is possible to progress while accurately understanding the calculation mechanics. Moment of inertia of area is a value that is identified by the cross-sectional shape of the part. This is used frequently for estimating the amount of deflection due to the bending moment or injection pressure. The moment of inertia of area changes depending only on the cross-sectional shape of a part. Therefore, it has no relationship with the material. For example, if the cross-sectional shape is the same, the value of the moment of inertia of area is the same whether the material is a non-heat treated steel, tempered steel, or even wood. The definition of moment of inertia of area according to mechanics is as follows. "When a cross-section is divided into an infinite number of differential areas δA and the distance from one axis X is a taken as Y, the moment of inertia of area is the sum over the entire area the product of the differential area and the square of the distance".

시 한번 이해하여 계산 기법을 정확하게 이해하면서 진행하자. 단면 2차 모멘트는 부품의 단면형상으로 식별되는 값이다. 이것은 굽힘 모멘트 또는 사출압력에 의한 변형량을 추정하는데 자주 사용된다. 단면 2차 모멘트는 오직 단면형상에 따라 변동한다. 따라서 재료와 관계가 없다. 예를 들어, 단면형상이 동일하다면, "단면 2차 모멘트"의 값은 비 열처리 스틸이건 열처리 스틸 또는 목재인지 여부에 관계없이 동일하다. 역학적 단면 2차 모멘트의 정의는 다음과 같다. "단면을 무한 수의 미분 영역 δA로 나누고, 한 축 X와의 거리를 Y로 취하면 단면 2차 모멘트는 전체 면적에서 미분 영역의 곱과 거리의 제곱 ".

This can be expressed in the form of an equation as follows.

다음과 같이 방정식의 형태로 표현 될 수 있다.

Cross-section second order moment is -

단면 2차 모멘트는

$$I = \Sigma Y^2 \times dA \quad \text{(unit is mm}^4 \text{ or m}^4\text{)}$$

$$I = \Sigma Y^2 \times dA$$

(단위 : mm^4 또는 m^4)

The moment of inertia of a area is usually represented by the symbol I as a matter of custom. In general, the strength against bending becomes larger as the moment of inertia of a area becomes larger. If the cross-sectional shape is rectangular or circular, the basic equation for calculation becomes clear as is shown in Table 1.

단면 2차 모멘트는 일반적으로 기호 I로 표시된다. 일반적으로, 굽힘에 대한 강도는 단면 2차 모멘트가 커질수록 커진다. 단면형상이 직사각형 또는 원형인 경우, 계산을 위한 기본 방정식이 표 1에 표시된 것처럼 명확해진다.

Cross-section shape	Moment of inertia of area	Modulus of section Z
Rectangle 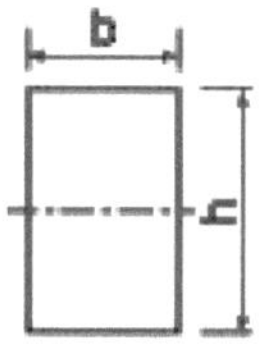	$I=\frac{bh^3}{12}$	$Z=\frac{bh^2}{6}$
Square 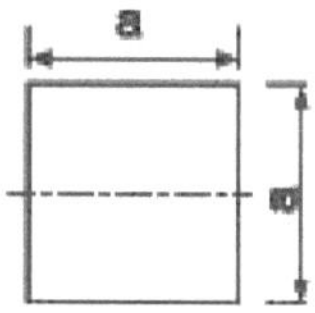	$I=\frac{a^4}{12}$	$Z=\frac{a^3}{6}$

Cross-section shape	Moment of inertia of area	Modulus of section Z
Circle 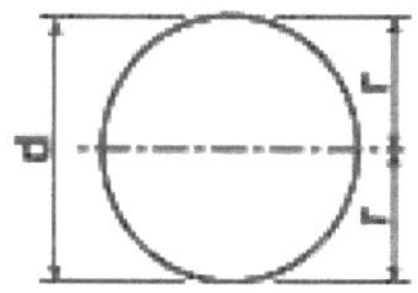	$I=\frac{\pi d^4}{64}=\frac{\pi r^4}{4}$	$Z=\frac{\pi d^3}{32}=\frac{\pi r^3}{4}$
Hollow circle 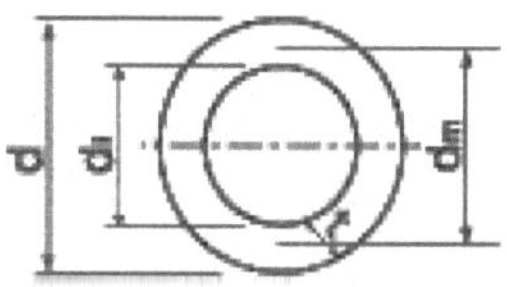	$t=\frac{\pi(d^4-d_I^4)}{64}$ Thin wall $I \fallingdotseq \frac{\pi}{8}t \cdot d_m^3$	$Z=\frac{\pi(d^4-d_I^4)}{32d}$ Thin wall $Z \fallingdotseq \frac{\pi}{4}t \cdot d_m^2$

G 009 Section Modulus of Mold Components

G-009 금형 부품의 단면 계수

The section modulus of mold components which is very important for predicting the bending stress is explained here.

At the time of calculating the strength against deformation or bending of a mold for plastic injection molding, the term "section modulus" appears very frequently. Let us understand how the "section modulus" is similar to the "moment of inertia of area" so as to carry out the mechanical calculations while understanding it more accurately.

The "section modulus" is a numerical value that is determined by the cross-sectional shape of the part. In that sense, it is similar to "moment of inertia of area".

The "section modulus" varies depending only on the cross-sectional shape of the part. Therefore, it has no relationship whatsoever with the material of the part. For example, if the cross-sectional shape is the same, the value of the "section modulus" will be the same irrespective of whether the material is non-heat treated steel, tempered steel, or even wood. The definition of section modulus according to mechanics is as follows. "A section modulus is the value of the "moment of inertia of area" related to the neutral axis of the cross-section of a beam multiplied by the distance from the neutral axis to the outer surface". Therefore, the relationship between the section modulus Z and the moment of inertia of area I is as expressed by the following equation.

굽힘 응력을 예측하는 데 매우 중요한 금형 부품의 단면 계수를 설명한다.

플라스틱 사출금형의 변형이나 굽힘에 대한 강도를 계산할 때, "단면 계수"라는 용어는 매우 자주 나타난다. "단면 계수"가 "단면 2차 모멘트"와 어떻게 유사한지를 이해하여 기계적 계산을 보다 정확하게 이해하면서 수행하자.

"단면 계수"는 부품의 단면형상에 의해 결정되는 값이다. 그런 의미에서 단면 계수는 "단면 2차 모멘트"와 유사하다.

"단면 계수"는 부품의 단면형상에 따라 달라진다. 따라서 부품의 재질과 아무런 관련이 없다. 예를 들어, 단면형상이 동일하다면, "단면 계수"의 값은 비 열처리 스틸이건 열처리 스틸 또는 목재인지 여부에 관계없이 동일하다. 역학적 단면 계수의 정의는 다음과 같다. "단면 계수는 중립축에서 외면까지의 거리를 곱한 보의 단면 중립축과 관련된 "단면 2차 모멘트"값이다." 따라서, 단면 계수 Z와 면적 I의 단면 2차 모멘트와의 관계는 다음 식과 같다.

Section modulus -

$$Z = \frac{I}{y} \text{ (unit is mm}^3 \text{ or m}^3\text{)}$$

Where, I is the moment of inertia of area and, y is the distance form the neutral axis to the outer surface.

The symbol Z is used customarily for the section modulus. In general, as the section modulus becomes larger, the strength against bending also becomes larger. Regarding bending, the maximum bending stress σ acting on the outer surface of the part can be calculated using the following equation.

$$\sigma = \frac{M}{Z}$$

Where, σ is the maximum bending stress (kgf/cm^2), and M is the maximum bending moment (kgf • m)

단면 계수 -

$$Z = \frac{I}{y} \text{(단위는 mm}^3 \text{ 또는 m}^3\text{)}$$

여기서, I는 단면 2차 모멘트이고, y는 중립축으로부터 외부 표면까지의 거리이다.

기호 Z는 일반적으로 단면 계수로 사용된다. 일반적으로 단면 계수가 커짐에 따라 굽힘 강도도 커진다. 굽힘에 대하여, 부품의 외부 표면에 작용하는 최대 굽힘 응력 σ는 다음 식을 사용하여 계산할 수 있다.

$$\sigma = \frac{M}{Z}$$

여기서, σ는 최대 굽힘 응력 (kgf/cm^2), M은 최대 굽힘 모멘트 (kgf·m)

G 010 Thermal Expansion of Mold Components

G-010 금형 부품의 열팽창

The basic knowledge about the thermal expansion of components in the molds for plastic injection molding is discussed below.

In the case of a mold for plastic injection molding, in order to maintain the appropriate cavity surface temperature, it is maintained at a temperature of 30 to 150℃. On the other hand, molten plastic flows into the sprue, runner, and cavity, which receive heat from the

플라스틱 사출금형의 열팽창에 관한 기본적인 지식은 아래에 설명되어 있다.

사출금형에서 적절한 캐비티 표면 온도를 유지하기 위해서 30 ~ 150℃의 온도가 유지된다. 반면 용융 된 플라스틱 수지는 180 ~ 300℃ 범위의 온도에서 스프루, 런너 및 캐비티로 흘러 들어간다.

plastic at temperatures in the range of 180 to 300℃. Metals generally undergo thermal expansion when the temperature rises. Therefore, even the constituent parts of a mold for plastic injection molding undergo thermal expansion. Thermal expansion can disrupt the mating between the guide post and guide bush, or can cause bad movement of the slide core, or can enlarge the dimensions of the core pins.

The basic changes in the dimensions due to thermal expansion can be calculated using the following equation.

$$\lambda = \alpha \cdot l_0 \cdot (t - t_0)$$

Where, λ is the expansion (in ㎜) in the dimensions expected to thermally expand, α is the linear thermal expansion coefficient (㎜/㎜) of the metal, l_0 is the initial length (㎜), t is the temperature after heating and t_0 is the initial temperature (℃)

The linear thermal expansion coefficients for typical metallic material used in molds are given below.

금속은 온도가 상승 할 때 일반적으로 열팽창 한다. 따라서 플라스틱 사출금형의 구성 부품조차도 열팽창을 겪게 된다. 열팽창은 가이드 핀과 가이드 부시 사이의 결합을 방해하거나 슬라이드 코어의 불량 동작을 초래하거나 코어 핀의 크기를 확대시킬 수 있다.

열팽창에 의한 치수의 기본적인 변화는 다음 방정식을 사용하여 계산할 수 있다.

$$\lambda = \alpha \cdot l_0 \cdot (t - t_0)$$

여기서, λ는 열팽창 할 것으로 예상되는 치수(㎜), α는 금속의 열팽창 계수(㎜/㎜), l_0는 초기 길이(㎜), t는 가열 후 온도, 그리고 t_0 초기 온도(℃), 이다.

금형에 사용되는 일반적인 금속재료의 열팽창 계수는 다음과 같다.

Material	α Linear thermal expansion coefficient (㎜/㎜)
S50C	11.7×10^{-6}
SKD11	11.7×10^{-6}
Pre-hardened steel (SCM440 series)	11.5×10^{-6}
18-8 Stainless steel	$(17\sim18) \times 10^{-6}$
36% Nickel steel	0.9×10^{-6}
Super duralumin	23.4×10^{-6}
Brass	$(18\sim23) \times 10^{-6}$
Copper	16.5×10^{-6}

G 011 Thermal Expansion of Mold Components(Example of Calculation)

Let us see a case study about the thermal expansion of mold components described in the last course.

Question:

The overall length of a core pin prepared in a 20℃ machine formation room was 30.52 ㎜. When this core pin is heated to 150℃, how long is the amount of thermal expansion? The material of the core pin is pre-hardened steel of the SCM440 series.

Sample Answer:

The equation for calculating the thermal expansion of metals is the following. Let us substitute the values in this equation.

$$\lambda = \alpha \cdot l_0 \cdot (t - t_0)$$

Where λ is the elongation (㎜) of the core pin due to thermal expansion and α is the linear thermal expansion coefficient of the metal (㎜/㎜).

In the case of a pre-hardened steel of the SCM440 series,

α = 11.5×10^{-6}㎜/㎜, l_0 is the initial length of the core pin (㎜),

l_0 = 30.52 ㎜, t is the initial temperature (℃),

t = 20℃, and t_0 is the temperature after heating (℃),

t_0 = 150℃

G-011 금형 부품의 열팽창 (계산 예)

마지막 과정에서 설명 된 금형 부품의 열팽창에 대한 사례 연구를 보기로 하자.

질문 :

20℃ 온도의 작업실에 준비된 코어 핀의 전체 길이는 30.52 ㎜ 이었다. 이 코어 핀이 150℃로 가열하면, 열팽창의 양은 얼마나 되나? 코어 핀의 재질은 SCM440 계열의 프리하든강이다.

답안 :

금속의 열팽창을 계산하는 공식은 다음과 같다. 우리는 이 방정식에 값을 대입 할 것이다.

$$\lambda = \alpha \cdot l_0 \cdot (t - t_0)$$

λ는 열팽창으로 인한 코어 핀의 길이(㎜)이며, α는 금속의 열팽창 계수 (㎜/㎜)이다.

SCM440 계열의 프리하든강의 경우, α = 11.5×10^{-6} ㎜/㎜, l_0는, 코어 핀(㎜)의 초기 길이, l_0=30.52 ㎜, t는 초기 온도(℃)이다. t= 20℃, 그리고 t_0는 가열 후 온도(℃)이며, t_0=150℃.

Therefore, $\lambda = 11.5 \times 10^{-6} \times 30.52 \times (150-20) = 0.0456$ (㎜).

따라서 $\lambda = 11.5 \times 10^{-6} \times 30.52 \times (150-20) = 0.0456$ (㎜).

G 012 Basic Dynamics of Injection Mold (Kinetic energy during mold opening and closing)

G-012 사출 금형의 기본 역학 (금형 개폐 중 운동 에너지)

Let us discuss the kinetic energy at the time of opening and closing the moving half mold after an injection mold has been installed in the injection molding machine. (Fig. 1)

사출성형기에 금형을 설치 한 후 금형을 열고 닫을 때의 운동 에너지에 대해 알아보자. (그림1참조)

Fig. 1

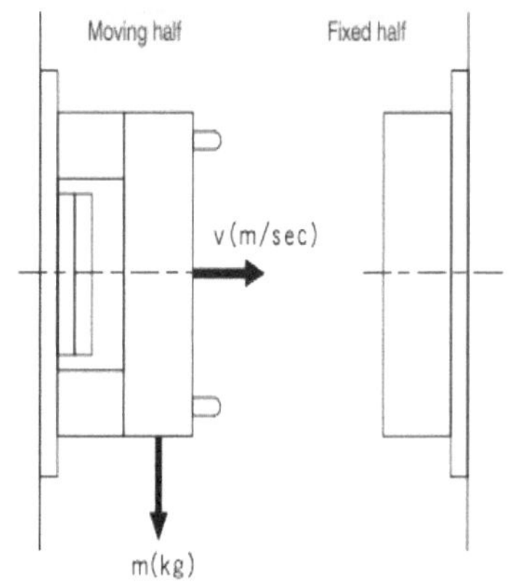

Kinetic energy is required when an object moves at a certain velocity. When the amount of kinetic energy becomes large, in order to stop the mold, a brake has to be applied that can withstand an equivalent or higher amount of energy. In addition, if the mold clamping is not done in the state in which the kinetic energy has been reduced sufficiently, there is the danger of the mold getting damaged or broken.

물체가 소정의 속도로 이동하려면 운동 에너지가 필요하다. 운동 에너지의 양이 커지면, 금형을 정지시키기 위한 제동 에너지도 운동 에너지 이상을 견딜 수 있게 적용되어야한다. 또한, 운동 에너지가 충분히 감소 된 상태에서 형체가 수행되지 않으면 금형이 손상되거나 파손될 위험이 있다.

The equation for calculating the kinetic energy is the following.

운동 에너지를 계산하기 위한 공식은 다음과 같다.

$E = 1/2mv^2$, where, E: Kinetic energy ($kg \cdot m^2 / sec^2$)
m : Mass(kg) v : Velocity(m/sec)

$E=1/2mv^2$, E : 운동에너지 ($kg{\cdot}m^2/sec^2$)
m : 질량 (kg) V : 속도 (m/sec)

The following points are evident from this equation

1) The kinetic energy E is proportional to the mass M on the movable side.
2) The kinetic energy E is proportional to the square of the movement velocity V on the movable side.

다음 사항이 방정식에 뚜렷이 나타나 있다.

1) 운동 에너지 E는 가동측의 질량 M에 비례한다.
2) 운동 에너지 E는 가동측의 이동속도 (V)의 제곱에 비례한다.

Therefore, as the mass of the moving half becomes large, even the kinetic energy becomes large proportionally. If the speed of mold opening and closing is made higher, even the energy increases proportional to the square of the speed.

Therefore, in the case of a mold in which a large number of core pins with ejector structure or pillbox structure have been placed, or a mold having abnormally shaped parting surface, at the time of mold clamping, if the slowing down (speed reduction) is not made sufficiently, since the mold may break due to the kinetic energy, it is necessary to take sufficient care.

따라서 이동측 금형의 질량이 커짐에 따라 운동 에너지도 비례하여 커진다. 금형의 개폐속도를 높게 설정하면 에너지도 속도의 제곱에 비례하여 증가한다.

따라서 이젝터 구조 또는 필박스 구조를 갖는 다수의 코어 핀이 배치 된 금형 또는 파팅면이 비정상적인 금형을 형체 할 때, 감속이 충분히 이루어지지 않으면 운동 에너지로 인해 금형이 파손될 수 있으므로 충분한 주의가 필요하다.

G 013 Basic Dynamics of Injection Mold (Force at the time of mold opening and closing)

G-013 사출 금형의 기본 역학 (금형 개폐 시의 힘)

This time we discuss the force required at the time of opening and closing the moving half mold after an injection mold has been installed in the injection molding machine.

Force is required to move an object. Force can be calculated using one of the famous Newton's laws of motion.

이번에는 사출성형기에 금형을 설치 한 후 이동측 금형을 개폐 할 때 필요한 힘에 대해 설명한다.

물체를 이동시키기 위해서 힘이 필요하다. 힘은 뉴튼의 운동 법칙 중의 하나를 이용하여 계산할 수 있다.

Newton's Law of Motion:
F = m.a. where, F = Force (kg.m/sec^2 or N),
m = Mass (kg) a = Acceleration (m/sec^2)

뉴턴의 운동 법칙 :
F = m.a 여기서, F = 힘(또는 N),
m = 질량(kg) a = 가속도(m/sec^2)

The point that should be noted here is that a is "acceleration" and not "velocity".

According to Newton's laws of motion:

1) Force is proportional to the mass.
2) Force is proportional to the acceleration.

Therefore, the force acting at the time of opening or closing a mold can be said to be proportional to the mass on the movable side and the acceleration.

여기에서 주목해야 할 점은 "속도"가 아닌 "가속도"이다.

뉴턴의 운동 법칙에 있어서 :

1) Force는 질량에 비례한다.
2) Force는 가속도에 비례한다.

따라서 금형의 열림과 닫힘 시에 작용하는 힘은 가동측 금형의 질량과 가속도에 비례한다.

G 014 Basic Dynamics of Injection Mold (Fluid Dynamics)

In an injection mold, molten plastic is injected into a mold, cooled to solidify it, and then the molded product is taken out. It is necessary to apply pressure in order to inject molten plastic.

How much pressure is required can be predicted in advance using calculations of fluid dynamics.

Bernoulli's Theorem:

$$v^2/2g + p/\gamma + h - H, \text{ where}$$

- v : Velocity of the fluid (m/s)
- g : Gravitational acceleration ($9.8 m/sec^2$)
- p : Pressure of the fluid (kgf/m^2)
- γ : (Gamma) Specific gravity of the fluid (kgf/m^3)
- h : Position head (m)
- H : Total head (m)

The item $v^2/2g$ is called the velocity head, and the item p/γ is called the pressure head.

Actually, molten plastic has some viscosity (resistance to flow), and since the viscosity changes upon time (as the plastic cools, the viscosity increases making it more difficult to flow), the equations of calculations of fluid dynamics become very difficult.

In this issue, we explain the "Bernoulli's theorem" which is the most fundamental principle for discussing the general state of fluids.

G-014 사출 금형의 기본 역학 (유체 역학)

사출성형에는 용융수지를 금형에 주입하고, 냉각하여 고화시킨 후, 성형품을 꺼낸다. 용융된 수지를 주입하기 위해서 압력을 가할 필요가 있다.

얼마나 높은 압력이 필요할지는 유체 역학의 계산을 사용하여 미리 예측할 수 있다.

베르누이의 정리 :

$$v^2/2g + p/\gamma + h = H$$

여기서

- v : 유체의 속도
- g : 중력 가속도
- p : 유체의 압력
- γ : 유체의 비중
- h : 위치 헤드
- H : 총 수두

$v^2/2g$는 수두 속도라고하고, p/γ는 압력 수두라고 한다.

실제로 용융 된 수지는 약간의 점도(유동 저항)를 가지며 시간에 따라 점도가 변하기 때문에(수지가 냉각되면서 점도가 증가하여 유동하기가 어려워 짐) 유체 역학적으로 계산하기가 매우 어렵다.

이번 단원에서 유체의 일반적인 상태를 논의하기 위한 가장 기본적인 원리인 "베르누이 정리"에 대해 설명한다.

In other words, in a fluid, the total energy is determined by the amount of energy based on the velocity, the amount of energy based on the pressure, and the amount position energy.

Since it is difficult to apply the Bernoulli's equation as it is when there are changes in the viscosity or compressibility of the fluid, fluid velocity, it cannot be applied as it is for actual plastic injection molding. However, this is a very important basic equation when considering the principles.

다시 말해서 유체에서 총 에너지는 속도를 기반으로 한 에너지 양, 압력을 기반으로 한 에너지 양 및 위치 에너지 양에 의해 결정된다.

유체의 점도나 압축성, 유체 속도에 변화가 있을 때 그대로 베르누이 방정식을 적용하기가 어렵기 때문에 실제 플라스틱 사출성형의 경우에 적용 할 수 없다. 그러나 이것은 원칙을 고려할 때 매우 중요한 기본 방정식이다.

G 015 Basic Dynamics of Injection Molds (Flow Volume Equation)

G-015 사출 금형의 기본 역학 (유체적 식)

In injection molding, the molten plastic (fluid) flows through sprues and runners and gets filled inside the cavity. This phenomenon is expressed schematically as a fluid with viscosity flowing in a pipe.

When a viscous fluid flows through a pipe having a certain cross-sectional area, there is a basic equation of calculation called the "Flow Volume Equation" that expresses the basic state.

Flow Volume Equation:

$$Q = AV, \text{ where,}$$

Q : Flow rate (discharge rate) (m^3/sec)

a : Cross-sectional area of the flow path (m^3)

v : Velocity of fluid flow (m/sec)

사출성형에서 용융수지는 스프루와 런너를 통과하여 캐비티 내부에 충전된다. 이 현상은 파이프에 점도를 가진 유체로서 도식적으로 표현할 수 있다.

점성유체가 특정 단면적을 갖는 파이프를 통해 흐를 때, 기본 상태를 표현하는 "유량 방정식"이라는 기본 계산식이 있다.

유량 방정식 :

$$Q = AV, \text{ 여기서,}$$

Q : 유량(토출 속도) (m^3/sec)

A : 유로 단면적 (m^3)

V : 유체 흐름의 속도 (m/sec)

The volume Q of fluid flowing through the pipe per unit time can be obtained by multiplying the cross-sectional area of the flow path (pipe) and the flow speed of the fluid.

단위 시간당 파이프를 통해 흐르는 유체의 체적 Q는 유로 (파이프)의 단면적과 유체의 유속을 곱하여 구할 수 있다.

It can be understood from this equation that in order to increase the flow rate, either the cross-sectional area a of the flow path has to be increased or the flow speed of the fluid has to be increased.

이 식으로부터, 유량을 증가시키기 위해서 유로 단면적(a)을 크게 하거나, 유체의 유속을 증가시켜야 한다는 것을 알 수 있다.

In addition, when the flow rate is constant, the following equation becomes valid.

또한 유량이 일정한 경우에는 다음식이 유효해진다.

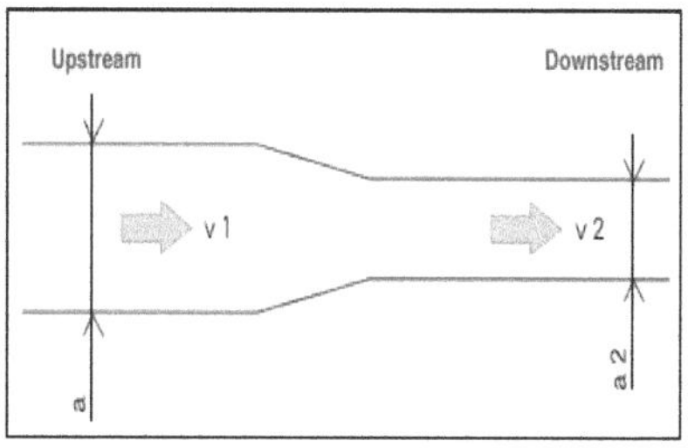

a1.v1 = a2.v2, where,

a1 : Cross-sectional area of the upstream flow path

a2 : Cross-sectional area of the downstream flow path

v1 : Upstream flow speed

v2 : Downstream flow speed

a1.v1 = a2.v2, 여기서,

a1 : 상류 유로의 단면적

a2 : 하류 유로의 단면적

v1 : 상류의 유속

V2 : 하류의 유속

In other words, when the flow rate is constant, the fluid flows slowly when the cross-sectional area of the flow path is large, and the fluid flow speed becomes fast when the cross-sectional area of the flow path becomes small.

즉, 유량이 일정한 경우, 유로의 단면적이 클 때에는 유체가 천천히 흐르고, 유로의 단면적이 작아지면 유체의 유속은 빨라진다.

Therefore, when the molten plastic flows inside a thin runner or gate, the flow speed becomes high.

따라서 용융수지가 얇은 런너 또는 게이트 내부로 흐를 때 유속이 빨라진다.

However, since the viscosity of the molten plastic changes with time, strictly speaking, the above equation will have to be corrected using a variety of factors.

그러나 용융수지의 점도는 시간에 따라 변하기 때문에 엄밀히 말하면 위의 방정식은 다양한 요인을 사용하여 보정해야 한다.

G 016 On the Number of Cavities in a Mold

G-016 금형의 캐비티 수

In plastic injection molding, the number of items that can be molded in one shot using one mold is called the number of cavities. The number of cavities can be from 1 to a reported maximum of 216.

Since the number of items molded in one shot becomes large as the number of cavities becomes large, it is possible to reduce the cost of production per item. However, as the number of cavities becomes larger, there is a trend of the quality fluctuations becoming large such as the dimensions, weight, external appearance, etc. of the molded item of the different cavities. It is necessary to determine whether to make the number of cavities large or small upon studying comprehensively the advantages and disadvantage.

In general, the following trend is seen in the number of cavities depending on the type of mold.

플라스틱 사출성형에서 하나의 금형을 사용하여 한 번에 성형 할 수 있는 제품의 수를 캐비티 개수라고 한다. 캐비티의 수는 1개에서부터 최대 216개까지 가능하다.

캐비티 수가 많아지면 한 번의 성형으로 성형되는 제품의 수가 많아지므로 제품 당 생산 비용을 줄일 수 있다. 그러나 캐비티의 수가 많아지면 다른 캐비티의 성형품 치수, 중량, 외관 등의 품질 변동이 커지는 경향이 있다. 장점과 단점을 종합적으로 분석하여 캐비티의 수를 많이 할지 또는 적게 할지 결정할 필요가 있다.

일반적으로 캐비티의 종류에 따라 다음과 같은 경향이 캐비티의 수에 나타난다.

○ **Large molded items ... Basically one cavity**

- Reason : Since the mold is large, it is physically difficult to have two or more cavities.

○ **대형 성형품 ... 기본적으로 하나의 캐비티**

- 이유 : 캐비티가 크기 때문에 두 개 이상의 캐비티는 물리적으로 어렵다.

○ **Medium to small molded items of automobiles, home electrical appliances, etc. ... About 4 to 8 cavities**

- Reason : Considering the relationship between the total production quantity and quality fluctuations, about this number of cavities is found to make the risk low by experience.

○ **자동차, 가정용 전기 제품 등의 중소형 성형품 ... 약 4 ~ 8 개의 캐비티**

- 이유 : 총 생산량과 품질 변동 사이의 관계를 고려해 볼 때, 캐비티의 수는 경험상 리스크를 낮추는 것으로 밝혀졌다.

○ **Electronic components, precision components ... 2 or 4 cavities**

- Reason : The number of cavities is made small considering the quality fluctuations of the molded item.

○ **전자 부품, 정밀 부품 ... 2 또는 4 개의 캐비티**

- 이유 : 성형품의 품질 변동을 고려하여 캐비티의 수를 줄여야 한다.

○ **Food containers, etc. ... 24 to 64 cavities, or more (hot runner specifications)**

- Reason : Since a production quantity of several million or more can be forecast, the number of cavities is made large thereby reducing the production cost.

○ **식품 용기, 기타 ... 24 ~ 64 개 이상의 캐비티 (핫런너 사양)**

- 원인 : 수백만 개 이상의 생산량을 예측할 수 있기 때문에 캐비티 수를 크게 하여 생산 원가를 절감한다.

○ **Molds of new products ... 1 cavity**

- Reason : Since the sales of a new product in the market cannot be forecast, initially it is produced using a mold having one cavity, and when the sales becomes brisk, a mold with increased number of cavities is prepared to meet the larger demand.

○ **신제품 금형 ... 1개의 공동**

- 이유 : 시장에서 신상품의 판매를 예측할 수 없으므로 처음에는 캐비티가 하나 인 금형을 사용하여 생산되며 판매량이 많아지면 캐비티 수가 증가한 금형이 더 큰 수요를 충족 할 수 있도록 준비한다.

G 017 Procedure for Determining the External Dimensions of a Cavity

G-017 캐비티의 외부 치수 결정을 위한 절차

How are the external dimensions of the cavity (nest in the fixed side) determined? In most cases, the reality is that the dimensions of a similar past mold are used as a reference, or the dimensions are determined by experience and intuition. If the correct procedure for determining the external dimensions is known, it is

캐비티의 외부 치수 (고정측 형판의 포켓)는 어떻게 결정되는가? 대부분의 경우 현재와 비슷한 과거 금형의 치수가 참조로 사용되거나 경험과 직관에 의해 결정된다. 외부 치수를 결정하는 올바른 절차가 알

possible to avoid the danger of accidents of the mold breaking due to the pressure of the plastic, and also to avoid the wastage of preparing an unnecessarily strong and large mold.

The correct procedure for determining the dimensions is explained below.

려져 있다면, 플라스틱의 압력으로 인해 금형이 파손될 위험을 피할 수 있으며, 불필요하게 강하고 큰 금형을 준비하는데 낭비되는 것을 방지 할 수 있다.

치수를 결정하는 정확한 절차가 아래에 설명되어 있다.

Step 1: Calculating the minimum wall thickness

1단계 : 최소 벽 두께 계산

A cavity is formed by carving a concave shape inside a block of steel material. Unless the thickness 'h' of the wall between the carved complementary shape of the molded product and the external shape of the steel material has a certain thickness, the mold may break or may become greatly deformed due to the filling pressure of the plastic. It is possible to obtain the recommended value of this thickness by theoretical calculations by applying the equations of the field of strength of materials.

The appropriate equation should be selected since the equation to be used differs depending on - (1) the external shape of the cavity (cubical or cylindrical), and (2) the structure of the cavity (unified or separated).

$$h = \sqrt[3]{\frac{12 \times p \times l^4 \times a}{384 \times E \times b \times \sigma_{\max}}}$$

The data to be substituted in the equation are determined considering the molding conditions, and the type of steel material, etc. The technique of a professional is also to assume a variety of cases such as the case of bad preconditions for calculation, a case of the best preconditions, etc., and to make a comparison of the results of the calculations. The minimum wall

캐비티는 강재 블록 내부를 오목한 모양으로 조각하여 형성된다. 조각 된 형상과 강재의 외부 형상 사이의 살의 두께 (h)가 일정한 두께를 갖지 않으면 사출압력으로 인해 금형이 파손되거나 크게 변형 될 수 있다. 재료 강도 분야의 방정식을 적용하여 이론적으로 계산하여 권장되는 두께 값을 얻을 수 있다.

사용되는 방정식에 따라 다르기 때문에 적절한 식을 선택해야 한다.

(1) 캐비티(입방체 또는 원통형)의 외부 형상

(2) 캐비티(단일 또는 분리)의 구조

$$h = \sqrt[3]{\frac{12 \times p \times l^4 \times a}{384 \times E \times b \times \sigma_{\max}}}$$

이 식에 대입 할 데이터는 성형 조건 및 강재의 종류 등을 고려하여 결정된다. 또한 전문가는 나쁜 조건과 좋은 조건 등의 여러 조건을 가정하여 계산하고, 그 계산의 결과를 비교한다. 이론에 기초한 최소 살 두께는 계산에 의해 얻어진

thickness based on theory is determined by taking into account a margin of safety for the value of h obtained by the calculations.

h의 값에 대한 안전 여유를 고려하여 결정된다.

Step 2: Calculating Cavity

If the external shape of the cavity is determined using the value of h obtained by calculations, at the time of fixing the cavity to the mold plate, etc., it may some times not be possible to obtain sufficient dimensions of flanges or to obtain sufficient space for drilling screw holes. In such cases, dimensions should be determined so that any one of these can be placed, and the final cavity dimensions are determined using even integer numbers that are round numbers (for example, 50 ㎜, 80 ㎜, etc.).

2단계 : 캐비티

계산에 의해 얻어진 h의 값을 이용하여 캐비티의 크기를 결정하면, 캐비티를 형판에 고정할 때, 충분한 크기의 플랜지가 확보되지 못할 수 있거나, 나사 구멍을 드릴작업하기에 충분한 공간을 확보하지 못할 경우가 가끔 발생할 수도 있다. 이 경우 치수 중 하나가 배치 될 수 있도록 치수를 결정해야하며, 최종 캐비티 치수는 정수를 사용해야 한다. (예 : 50 ㎜, 80 ㎜ 등)

[Fig. 1]

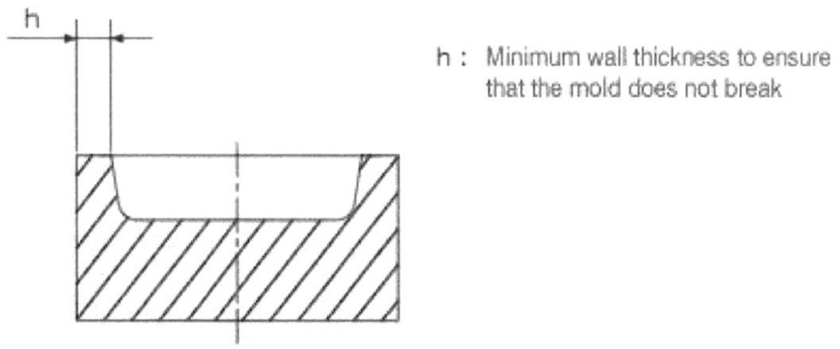

[Fig. 2]

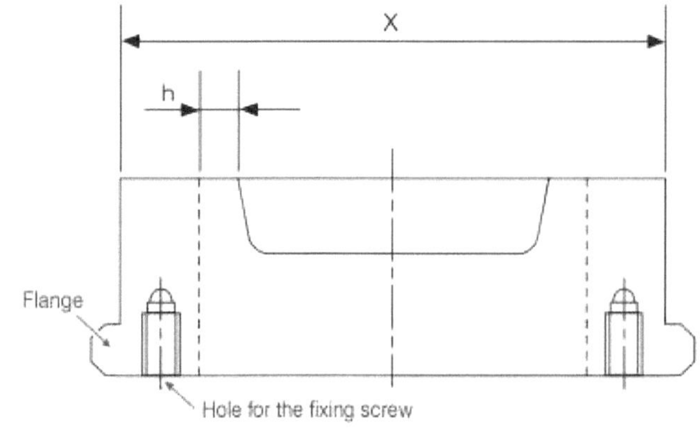

G 018 Calculation of the Deflection of the Moving half Cavity Plate

Do you have the experience of flash generated on the periphery of the parting surface of the molded product, or the height dimension of the molded product near the sprue becoming higher?

The shape of the moving half cavity plate that is the basis of calculations is shown in Fig. 1.

G-018 가동측 캐비티 플레이트의 변형량 계산

성형품의 파팅면 주변에서 플래시가 발생하거나 스프루 근처에서 성형품의 높이가 커졌던 경험을 가지고 있는가?

계산의 기초가 되는 가동측 캐비티 플레이트의 형상이 그림 1과 같다.

[Fig. 1]

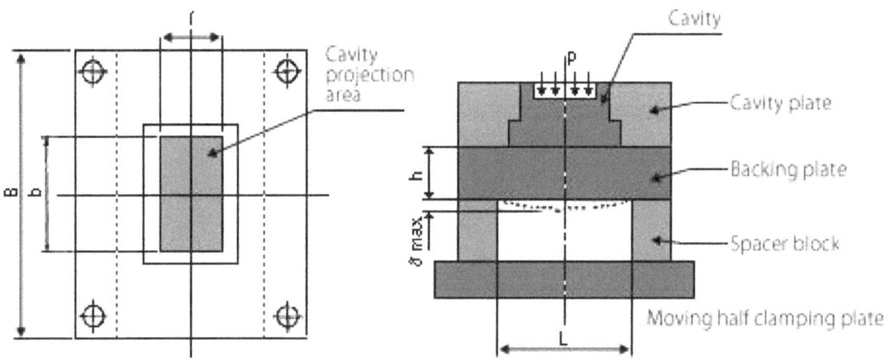

The maximum deflection δmax occurs along the center line of the cavity plate. The equation for calculating the deflection is as follows.

$$\delta max = \frac{5 \times p \times b \times L^4}{32 \times E \times B \times h^3}$$

B : Width of cavity plate (㎜)

b : Width (㎜) of the part receiving the cavity injection pressure p

L : Spacing (㎜) on the inside of the spacer block

p : Cavity internal injection pressure (kgf/cm^2)

h : Thickness of the backing plate (㎜)

E : modulus of longitudinal elasticity (Young's modulus) of the material (kgf/cm^2)

I : Length (㎜) of the part receiving the cavity internal injection pressure p

δmax : Maximum deflection (㎜) of the backing plate.

The important data for E (modulus of longitudinal elasticity) of the mold plate and p (cavity internal injection pressure) are given below.

최대 처짐 δmax는 캐비티 플레이트의 중앙 라인을 따라 발생하다. 처짐을 계산하는 공식은 다음과 같다.

$$\delta max = \frac{5 \times p \times b \times L^4}{32 \times E \times B \times h^3}$$

B : 캐비티 플레이트의 폭 (㎜)

b : 사출압력 P를 받는 캐비티 폭 (㎜)

L : 스페이스 블록의 내부 간격 (㎜)

P : 캐비티 내부의 사출압력 (kgf/cm^2)

h : 받침판 두께 (㎜)

E : 재료의 종탄성계수(Young's modulus) (kgf/cm^2)

l : 사출압력 p를 받는 캐비티 내부의 길이 (㎜)

δmax : 받침판의 최대 처짐 (㎜).

금형 플레이트의 E(종탄성계수)와 P(캐비티 내부의 사출압력)에 대한 중요한 데이터는 다음과 같다.

Type of cavity plate	**Value of E**
Material	E(kgf/cm^2)
S50C	210×10^4
Pre-hardened steel (SCM440 series)	230×10^4
Ultra duralumin	73×10^4

Cavity internal injection pressure	**(kgf/cm^2)**
Lower injection pressure	200 ~ 400
Higher injection pressure	400 ~ 600

The above equation for calculating the deflection is one for carrying out an approximate calculation. In actuality, since the pocket hole of the slide core and the holes for ejector pins have been formed in the cavity plate, and even the shape of the cavity is not uniform, it can be said that carrying out an accurate calculation of the deflection is actually very difficult. Therefore, the realistic method is to carry out the basic calculation using the approximate equation, and to correct the result to be on the safer side, or to consider factoring in a margin.

처짐 계산을 위한 위의 방정식은 대략적인 계산을 수행하기 위한 것이다. 실제로, 캐비티 플레이트 내에 슬라이드 코어의 포켓 구멍과 이젝터 핀 구멍이 있고, 캐비티 모양이 균일하지 않으면 정확한 변형 계산을 수행하는 것은 실제로 매우 어렵다. 따라서 현실적인 방법은 대략 방정식을 사용하여 기본 계산을 수행하고, 안전을 고려하여 결과를 수정하거나 여유를 고려해야 한다.

G 019 Preventing the Position Shift of Left-Right Asymmetric Cavities

G-019 좌우 비대칭 캐비티의 편심 방지

When the placement of the cavity is left-right asymmetric with respect to the sprue (see the example in Fig. 1), depending on the shape of the cavity, at the time of filling the molten plastic it is possible that a shift occurs in the relative positions of the cavity block and the core block, thereby causing changes in the dimensions of the molded product, or generation of burrs.

캐비티의 배치가 스프루에 대해 좌우 비대칭인 경우(그림 1의 예 참조)에 캐비티의 모양에 따라 용융수지를 충전할 때 시프트가 발생할 수 있다. 캐비티 블록과 코어 블록의 상대적인 위치에 의해 성형품의 치수가 변하거나 버가 발생하는 등의 문제가 있다.

Fig. 1

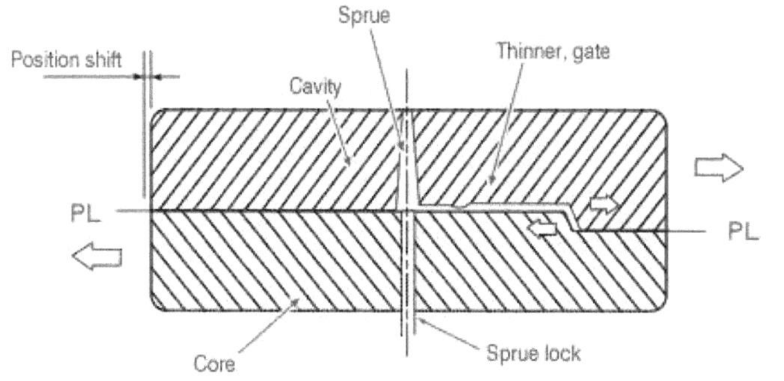

In order to prevent such problems, as is shown in Fig. 2, by placing a taper block set in a direction opposite to that of the cavity carving it is possible to prevent the position shifts.

이러한 문제를 방지하기 위해서 그림 2에 나타낸 것처럼 테이퍼 블록을 캐비티의 방향과 반대 방향으로 배치함으로써, 위치 이동을 방지할 수 있다.

Fig. 2

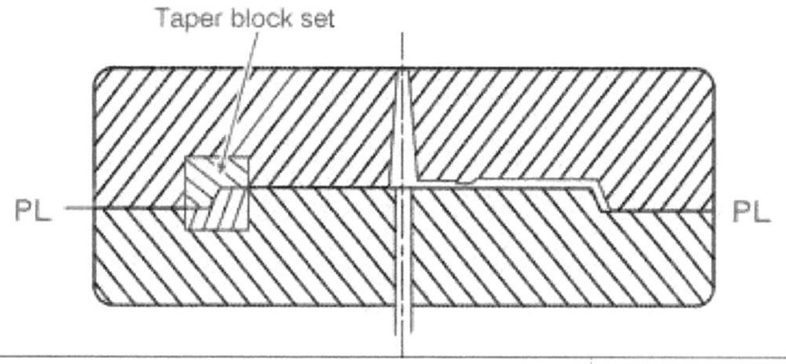

In the case of a single cavity mold or a test mold, a situation such as that shown in the example of Fig. 1 can occur. In such a situation, instead of concentrating completely on the preparation of the shape of the molded product, try to look at the overall structure of the mold from a different perspective from a distance, and remember the hints given here.

One can achieve a balanced design if one is not paying attention only to some parts but has the flexibility to view the whole.

단일 캐비티 금형 또는 테스트 금형의 경우에도 그림 1의 예에 나타낸 것과 같은 상황이 발생할 수 있다. 이러한 상황에서 성형품의 형상에 집중하는 대신 먼 곳에서 다른 관점으로 금형의 전체 구조를 살펴보고 여기에 제시된 힌트를 기억하자.

하나의 부품에만 주의를 기울이지 않고, 전체를 볼 수 있는 유연성을 가진다면 균형 잡힌 설계를 할 수 있다.

G 020 Method of Determining Cavity Dimensions (1)

G-020 캐비티 치수 결정 방법(1)

In the previous lesson we have explained the molding shrinkage ratio and in this lesson we describe a method of determining the cavity dimensions using the molding shrinkage ratio.

이전 강의에는 성형 수축률에 대해 설명했으며, 이 강의에서는 성형 수축률을 사용하여 캐비티 치수를 결정하는 방법을 설명한다.

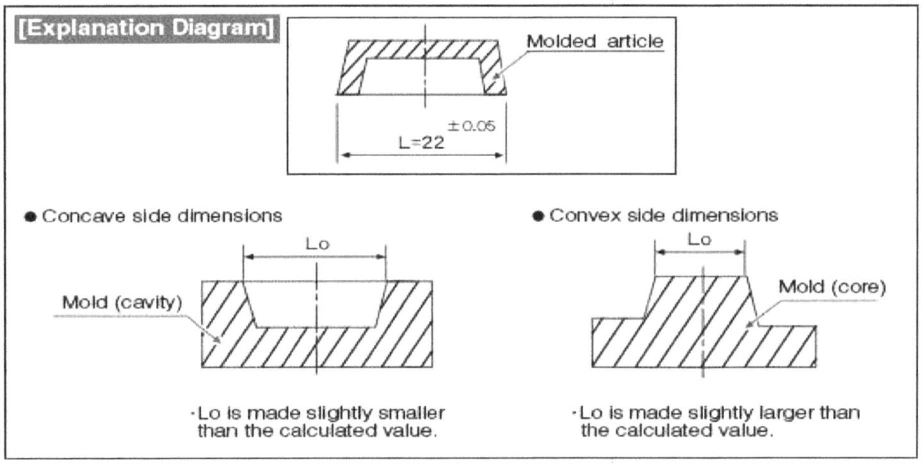

Firstly, we determine the molding shrinkage ratio α.
Example: α = 0.005

Next, we select the dimension of the molded article that is to be determined. (When the dimensional tolerance is a ± tolerance value)
Example: 22 ±0.05

Next, we calculate the cavity dimension that is to be obtained using Eqn. 2 below.

우선, 성형 수축률 α를 구한다.

예 : α = 0.005

다음은 성형품의 치수를 결정한다. (치수 공차가 ± 공차 값일 때)

예 : 22 ± 0.05

다음은 식 2를 이용하여 캐비티 치수를 계산한다.

$$L_0 = (1+\alpha) \times L \quad \dots \text{[Eqn. 2]}$$

Where,

L_0 : Cavity dimension (㎜)

L : Dimension of the molded article (㎜)

α : Molding shrinkage ratio

Example:

$$L_0 = (1+\alpha) \times L$$

$$= (1+0.005) \times 22$$

$$= 1.005 \times 22$$

$$= 22.11$$

Therefore, theoretically the target dimension of the cavity will be 22.11 ㎜.

In addition, slight adjustment is made in the calculated dimension of 22.11 ㎜ considering machine operability and mold correction after trial fabrication.

Although machining a mold component with a width of 22.11 ㎜ is of course possible with the present day machines, since the machining dimensions are specified in units of 0.01 ㎜ and since the machining cost will be high, if possible we would like to round the 0.01 ㎜ part of the dimension to an even number.

Therefore, we correct as 22.11 → 22.10 or 22.11 → 22.12. Of course it is possible to leave it as 22.11 if we want to machine as precisely as possible.

Finally, when the calculated dimension is on the concave side of the mold, in some cases, the adjustment of the dimension on the convex side is made one more time than in the case of the concave side. In the case of the concave side, it is also possible to make the dimension slightly smaller so that the mold can be corrected later on. In this case, the correction is made

$$L_0 = (1+\alpha) \times L \quad \dots \text{[식 2]}$$

여기서,

L_0 : 캐비티 치수 (㎜)

L : 성형품 치수 (㎜)

α : 성형 수축률

예 :

$$L_0 = (1+\alpha) \times L$$

$$= (1+0.005) \times 22$$

$$= 1.005 \times 22$$

$$= 22.11$$

이론적으로 캐비티의 목표 치수는 22.11 ㎜가 된다.

또한 시험 제작 후 성형기 작동과 금형 수정을 고려하여 계산 된 22.11 ㎜ 치수에서 약간의 조정이 이루어진다.

현재 기계로 22.11 ㎜ 너비의 금형 부품 가공이 딩연히 가능하지만, 가공 치수가 0.01 ㎜ 단위로 지정되어 있고, 기계 가공 비용이 높기 때문에 가능한 경우 0.01 ㎜ 부분을 짝수로 반올림한다.

따라서 22.11→22.10 또는 22.11→22.12로 수정한다. 가능한 한 정확하게 가공하려면 22.11로 남겨둘 수 있다.

최종적으로 계산 된 치수가 금형의 캐비티 상에 있을 때, 코어 치수의 조정은 캐비티의 경우보다 한번 더 이루어진다. 캐비티의 경우에 치수를 약간 작게 하여 나중에 금형을 수정할 수 있다. 이 경우 22.10 → 22.08 또는 22.12 → 22.08로 보정

as 22.10→22.08 or 22.12→ 22.08. The mold is prepared with a slightly larger dimension on the convex side. In this case, the correction is made as 22.10 → 22.12 or 22.12 → 22.14. These adjustments are not necessary if it is judged based on experience that the mold need not be corrected.

한다. 금형은 코어에서 약간 큰 치수로 준비한다. 이 경우, 22.10 → 22.12 또는 22.12 → 22.14로 보정된다. 이러한 조정은 경험에 근거하여 금형이 수정 될 필요가 없다고 판단되는 경우 필요하지 않다.

G 021 Method of Determining Cavity Dimensions (2)

G-021 캐비티 치수 결정 방법(2)

In the last lesson, while we explained a basic method of determining the cavity dimensions, in this lesson we describe an example of application of this method.

마지막 단원에는 캐비티 치수를 결정하는 기본 방법을 설명했지만 이번 강의에는 이 캐비티 치수 결정 방법에 대한 적용 예를 설명한다.

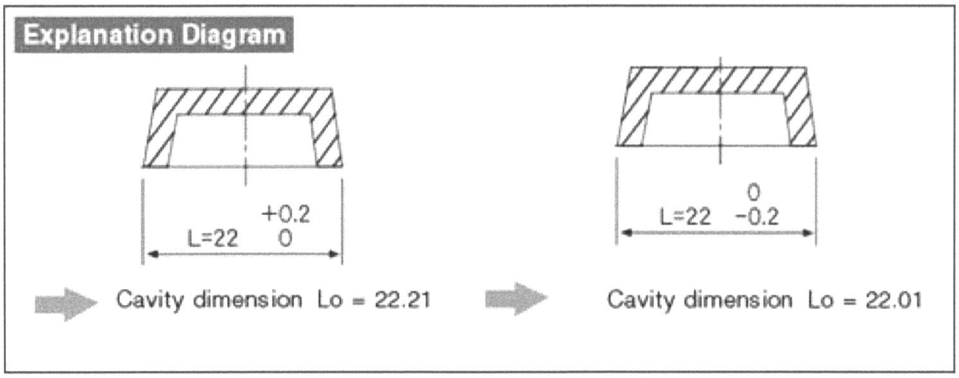

Application Example 1: When the dimensional tolerance is a single sided tolerance value (when it is not a ± tolerance value)

적용 사례 1 : 치수 공차가 단면 공차 값 일 때 (± 공차 값이 아닌 경우)

Let us consider the case when a dimension of the molded article is 22.

성형품의 치수가 22인 경우를 생각해 보자.

Example of Failure:

$$L_0 = (1+\alpha)\times L = (1+0.005)\times 22 = 1.005\times 22 = 22.11$$

실패의 예 :

$$L_0 = (1+\alpha)\times L = (1+0.005)\times 22 = 1.005\times 22 = 22.11$$

If the cavity is prepared based on this calculation result, since it can be considered that the probability of fluctuations in the dimensions of the molded article will be the same in the + direction as well as in the - direction, in case the shrinkage becomes larger than predicted, there is the danger that the - side tolerance will be exceeded.

In view of this, in the case of such single sided tolerance, the calculation of shrinkage is made aiming at the center of the width of tolerance.

계산 결과에 의해서 캐비티를 준비하면 성형물의 치수 변동 확률은 + 방향뿐만 아니라 - 방향에서도 동일하기 때문에 수축이 커지면 예측한 것보다 측면 허용 오차가 초과 될 위험이 있다.

이러한 관점에서 단면 공차의 경우 수축 계산이 공차 폭의 중심에 기준을 두어야 한다.

Appropriate Example:

$$L_0 = (1+\alpha)\times L = (1+0.005)\times(22+(0.2/2)) = 1.005\times 22.1 = 22.21$$

적절한 예 :

$$L_0 = (1+\alpha)\times L = (1+0.005)\times(22+(0.2/2)) = 1.005\times 22.1 = 22.21$$

Application Example 2: When the dimensional tolerance is a single sided tolerance value (when it is not a ± tolerance value)

적용 예 2 : 치수 공차가 단면 공차 값일 때 (±공차 값이 아닌 경우)

Let us consider the case when a dimension of the molded item is 22.

성형품의 치수가 22인 경우를 생각해 보자.

Example of Failure:

$$L_0 = (1+\alpha)\times L = (1+0.005)\times 22 = 1.005\times 22 = 22.11$$

실패의 예 :

$$L_0 = (1+\alpha)\times L = (1+0.005)\times 22 = 1.005\times 22 = 22.11$$

Even when the cavity is prepared based on this result of calculation, as in the above example, in case the shrinkage becomes smaller than predicted, there is the danger that the + side tolerance will be exceeded.
In view of this, even in the case, the calculation of shrinkage is made aiming at the center of the width of tolerance.

계산 결과에 의해 캐비티를 준비한 경우에도 상기 예와 같이 수축이 예측보다 작아지면 + 측의 공차를 초과 할 우려가 있다.

이러한 관점에서 수축의 계산은 공차 폭의 중심에 기준을 두어야 한다.

Appropriate Example:

$$L_0 = (1+\alpha)\times L = (1+0.005)\times(22-(0.2/2)) = 1.005\times 21.9 = 22.01$$

적절한 예 :

$$L_0 = (1+\alpha)\times L = (1+0.005)\times(22-(0.2/2)) = 1.005\times 21.9 = 22.01$$

In this manner, the cavity dimensions are determined appropriately for each dimensional tolerance of the molded article.

이러한 방식으로 캐비티 치수는 성형품의 각 치수 공차에 대해 적절히 결정한다.

G 022 Strength Calculation of the Thickness of the Side Wall of a Cavity (In the case of a rectangular cavity with bottom surface parting)

G-022 캐비티 측벽 살 두께의 강도 계산 (바닥면이 분리 된 직사각형 캐비티의 경우)

While in the last lesson we have explained the method of determining the external dimensions of a cavity (fixed side inserts), from this lesson onwards we explain the actual calculation of the strength of the thickness of the side wall of a cavity.

마지막 강의에서 캐비티의 외부 치수 (고정 된 측면 삽입물)를 결정하는 방법을 설명했지만 이 강의에서부터 캐비티의 측면 살 두께의 실제 강도를 계산한다.

This time we shall take the example of calculating the side wall thickness of a rectangular cavity whose bottom surface is parted.

The shape of the cavity in this example is shown in Fig.1.

이번에는 바닥면이 분리 된 직사각형 캐비티의 측면 살 두께를 계산하는 예제를 선정한다.

캐티비 형상이 그림 1에 나타나 있다.

[Fig.1]

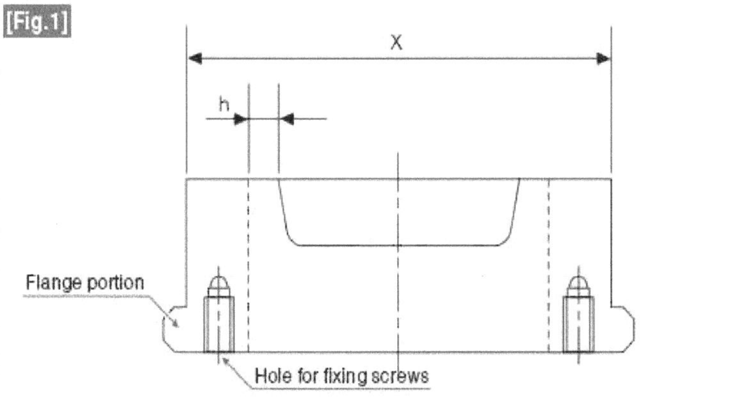

The thickness h of the side wall in Fig. 1 can be obtained using the following equation.

그림 1의 측면 살 두께 (h)는 다음 식을 사용하여 구할 수 있다.

$$h = \sqrt[3]{\frac{12 \cdot p \cdot l_4 \cdot a}{384 \cdot E \cdot b \cdot \delta_{max}}}$$

$$h = \sqrt[3]{\frac{12 \cdot p \cdot l_4 \cdot a}{384 \cdot E \cdot b \cdot \delta_{max}}}$$

Here,

h : Thickness of the cavity wall (㎜),

p : Cavity internal pressure (kgf / cm^2),

l : Inside length of the cavity (㎜),

a : Height of the side wall receiving the cavity internal pressure p (㎜),

b : Height of the cavity (㎜),

E : Modulus of elasticity (Young's modulus) (kgf / cm^2),

δmax : Maximum permissible strain(㎜)

여기서,

h : 캐비티 살의 두께 (㎜),

p : 캐비티 내압 (kgf/cm^2),

l : 캐비티의 내부 길이 (㎜)

a : 캐비티 내압을 반는 측면 살이 높이 (㎜),

b : 캐비티 높이 (㎜),

E : 탄성 계수(영률) (kgf／cm^2),

δmax : 최대 허용 변형률 (㎜),

Calculate the values of each of the above variables according to the units within parentheses ().

괄호 안에 있는 단위에 따라 위의 각 변수의 값을 계산하자.

The internal pressure p of the cavity takes on values in the range of 200 to 600 kgf/cm^2. The value of p varies depending on the plastic and the molding conditions. For example, the value will be about 400 kgf/cm^2in the case of ABS plastic, and will be about 600 kgf/cm^2 the case of PC plastic.

캐비티의 내부 압력 (p)은 200 ~ 600 kgf/cm^2 범위의 값이다. p의 값은 플라스틱 및 성형 조건에 따라 다르다. 예를 들어 ABS 플라스틱의 경우 약 400 kgf/cm^2이고, PC 플라스틱의 경우 약 600 kgf/cm^2이다.

l is the length of the periphery of the cavity,
a is the depth of the periphery of the cavity, and
b is the height of the cavity,
E is the modulus of elasticity of the steel material of the cavity.
See the data in some back numbers for the value.

l은 캐비티 주변의 길이,
a는 캐비티 주변의 깊이,
b는 캐비티의 높이,
E는 캐비티용 강재의 탄성 계수이다.
값은 뒷부분의 데이터를 참조하자.

δmax is the maximum permissible strain of the side wall. A value of 0.01 to 0.02 ㎜ is appropriate.

δmax는 측면 살의 최대 허용 변형률이다. 0.01 ~ 0.02 ㎜의 값이 적절하다.

The data substituted in the equation for calculation should be determined considering the molding conditions or the type of steel, etc.

계산식에서 대체되는 데이터는 성형 조건이나 강의 종류 등을 고려하여 결정되어야한다.

In the next lesson, we will carry out actual calculations for an example.

다음 강의에는 예제에 대한 실제 계산을 수행한다.

G 023 Selection of the Runner Cross-sectional Shape

G-023 런너 단면형상 선택

The runner is a path for making the molten plastic flow from the sprue to the molded product. The cross-sectional shape of the runner is selected depending on the size of the molded product, the type of plastic, the estimated molding conditions, etc. The

런너는 용융 수지를 스프루에서 성형품이 만들어지는 곳까지 흐르게 하는 통로이다. 런너의 단면형상은 성형품의 크기, 플라스틱의 종류, 예상되는 성형조건 등에 따라 선택

basic standards for selecting the runner cross-sectional shape are explained here.

The figure below shows some typical runner cross-sectional shapes.

된다. 런너 단면형상을 선택하기 위한 기본적인 표준 형상이 여기에 설명되어 있다.

아래 그림은 일반적인 런너 단면형상을 보여준다.

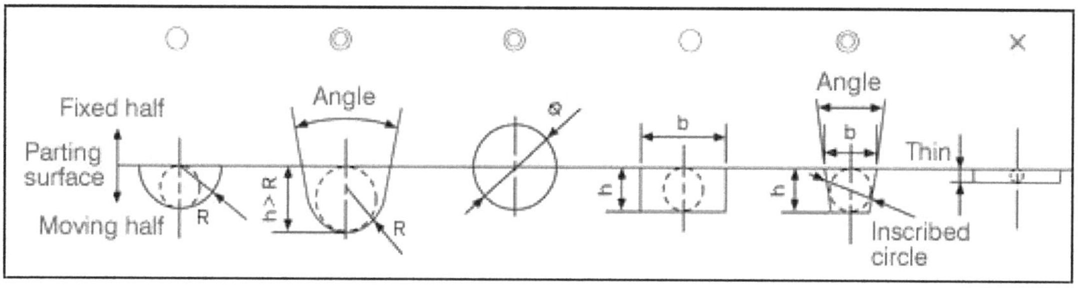

The runner is selected from the following three types.

1) The type carved on the moving half
2) The type carved on the fixed half
3) The type carved on both the fixed half and the moving half.

런너는 다음 3가지 형상 중에서 선택한다.

1) 가동측 형판에 가공
2) 고정측 형판에 가공
3) 고정측 형판과 가동측 형판 양쪽에 가공

The selection among these is made appropriately based on the restrictions imposed by the shape of the molded product, the mold parting position, etc.

The symbols in the figure have these meanings:

◎: Most suitable, ○: Suitable, ×: Not suitable

이들 중의 선택은 성형품의 형상, 성형품 취출 위치 등에 기초하여 적절하게 선택한다.

그림의 기호는 다음의 의미를 갖는다.:

◎ 최적, ○ 적당한, × 적합하지 않음

The most important role of the runner is to make the molten plastic flow in a condition in which no pressure loss is present. An unnecessarily large volume runner has the bad effects of increasing the scrap, worsening the material cost, lengthening the molding cycle, increasing the waste material, etc.

런너의 가장 중요한 역할은 압력 손실이 없는 상태로 용융 수지가 흐르게 하는 것이다. 불필요하게 큰 볼륨의 런너는 스크랩 증가, 재료비용 증가, 성형 사이클 타임 연장, 폐기물 증가 등의 나쁜 영향을 미친다.

An index of the efficiency of a runner is the diameter and area of the inscribed circle in the cross-sectional shape. If the area of the inscribed circle increases, the area for hot plastic to flow becomes wide, and hence the molten plastic can flow easily.

Consequently, a runner with a circular cross section is the most ideal one. However, since it will be necessary to carve the runner in both the fixed side and the movable side, the cost of manufacturing the mold increases. In order to solve these problems, a rectangular cross-sectional shape runner or a runner with a semicircular cross-sectional shape that is made still deeper is used.

It is desirable to provide an angle at the side surface of the runner for separating from the mold. In addition, the inner surface of the runner is polished to a smooth surface using a rubber grinder or by carrying out lapping, thereby preventing pressure loss.

런너 효율의 지표는 단면형상에서 내접원의 직경과 면적이다. 내접원의 면적이 커지면 고온 플라스틱의 유동 영역이 넓어 용융 수지가 쉽게 유동 할 수 있다.

결과적으로 원형 단면을 가진 런너가 가장 이상적이다. 그러나 고정측 및 가동측 모두에서 런너를 가공할 할 필요가 있기 때문에 금형 제조비용이 증가한다. 이러한 문제를 해결하기 위해 단면형상이 직사각형인 런너 또는 반원형의 런너가 사용된다.

금형으로부터 분리하기 위해 런너의 측면에 구배를 주는 것이 바람직하다. 또한, 고무 그라인더를 사용하거나 래핑(lapping)을 실시하여 런너의 내면을 매끄러운 표면으로 연마함으로써 압력손실을 방지해야 한다.

G 024 Hot Runner Technology

G-024 핫 런너 기술

A hot runner has the feature that it is possible to carry out injection molding without generating runner parts such as scrap, etc., by incorporating an electrically heated runner part in the mold.

Since hot runners make it possible to make the amount of scrap generated during the large scale manufacturing of plastic injection molded products extremely small, hot runners are a technology that make it possible to achieve great reductions in the cost of scrap processing.

핫 런너는 전기적으로 가열 되는 런너 성형용 부품을 금형에 결합하므로 런너와 같은 스크랩의 발생 없이 사출성형을 수행 할 수 있다는 특징이 있다.

핫 런너는 플라스틱 사출성형 제품의 대규모 생산 과정에서 발생하는 스크랩의 양을 극히 줄일 수 있으므로 핫 런너는 스크랩 처리 비용을 대폭 절감 할 수 있는 기술이다.

In addition, since the filling pressure loss that occurs in the case of cold runners becomes small in the case of hot runners, it is possible to suppress a fall in the filling pressure. Further, it is also possible to aim at shortening the cooling cycle. Hot runners have been used in a large number of cases, such as for the production of food containers, medical implements, automotive parts, etc.

Hot runner systems are broadly classified into those in which the mechanism part is a manifold part and those in which the mechanism part is a nozzle part. In addition, a controller is also required for carrying out the temperature control of the hot runner part.

Hot runners can be designed and built in-house within the company, or can be purchased from outside as systems. In general, the method of purchasing and applying a system available in the market is used most often. About 20 foreign companies are supplying such systems to the market, and all of them have their own technical features for the method of heating or for the heat transmission structure.

For example, the external heating method or the internal heating method can be used as the nozzle heating method. Further, various valve gate structures have been developed, such as the one in which the gate is opened and closed mechanically by force.

Since there are different types of hot runners that are used depending on the material to be molded and since molding is difficult in the case of some hot runners, it is necessary to sufficiently study in advance the type of plastic and the presence or absence of glass fibers, etc. During the design of the mold, it is necessary to propose the structure for incorporating a hot runner or manifold, the countermeasures against thermal expansion, the cooling structure, the maintenance structure, etc.

또한, 핫 런너의 경우 콜드 런너와 비교하여 사출압력손실이 작기 때문에 사출압력의 저하를 억제 할 수 있다. 더 나아가, 냉각 사이클 타임의 단축이 가능하다. 식품 용기, 의료 도구, 자동차 부품 등의 많은 경우에 사용되고 있다.

핫 런너 시스템은 매니폴드 구조와 노즐 구조로 크게 구별된다. 또한 핫 런너 부품의 온도 제어를 위하여 컨트롤러가 필요하다.

핫 런너는 사내에서 설계 또는 제작하거나 외부에서 시스템으로 구입할 수 있다. 일반적으로 시장에서 판매되는 시스템을 구매하고 적용하는 방법이 가장 많이 사용된다. 약 20여개의 외국 기업이 핫 런너 시스템을 공급하고 있으며, 이들 모두는 가열 방식이나 열 전달 구조에 대하여 자체 기술 특성을 가지고 있다.

예를 들면, 노즐 가열법으로 외부 가열 법 또는 내부 가열 법을 사용할 수 있다. 또한, 게이트가 힘에 의해 기계적으로 개폐되는 것과 같은 다양한 밸브 게이트 구조가 개발되었다.

성형되는 소재에 따라 사용되는 핫 런너의 종류가 다양하고 일부 핫 런너의 경우 성형이 어렵기 때문에 미리 플라스틱의 종류와 유리섬유 등의 유무를 충분히 연구 할 필요가 있다. 금형설계 시에는 핫 런너 또는 매니폴드를 금형과 통합하기 위한 구조, 열팽창 대책, 냉각 구조, 유지보수 구조 등을 제안해야 한다. 핫

The design is different from that of an ordinary mold, and it is necessary to thoroughly investigate the thermal calculations and the strength calculations. In addition, since recovering the initial investment for hot runners is difficult unless the production quantity of the molded product is large, it is necessary to verify the plan in detail as to how much planned production is to be made using that mold. In recent times, a trend can be seen in using valve gates or new types of hot runners in the development of molded products of super engineering plastics whose material costs are high. This technology is being considered to be used more and more frequently in the future as a means for making the costs of molding materials a minimum and for realizing molding that is in harmony with the environment.

런너 설계는 일반적인 금형의 설계와 다르며, 열 계산과 강도 계산을 철저히 할 필요가 있다. 또한 성형품의 생산량이 많은 경우가 아니면 핫 런너의 초기 투자 회수가 어렵기 때문에 해당 금형을 사용하여 얼마나 생산할 것인지에 대한 상세한 계획을 검증할 필요가 있다. 최근에는 재료비가 높은 슈퍼 엔지니어링 플라스틱 성형품 개발에 밸브 게이트 또는 새로운 종류의 핫 런너를 사용하는 추세이다. 이 기술은 성형 재료비용을 최소화하고 환경과 조화를 이루는 성형을 실현하는 수단으로 앞으로 점점 더 빈번하게 사용될 것으로 생각되고 있다.

G 025 Structure and Characteristics of Hot Runners

G-025 핫 런너의 구조 및 특징

Hot runners are a method of carrying out molding without generating scraps by heating and melting the runner part during plastic injection molding. Various types of structures have been realized for hot runners such as the method of heating or injecting. The major hot runner structures and their characteristics are as follows.

핫 런너란 플라스틱 사출성형 시에 런너 부를 가열 용해하여 스크랩을 발생시키지 않고 성형하는 방법이다. 핫 런너를 가열 또는 주입하는 방법과 같은 다양한 유형의 구조가 현재 사용된다. 주요 핫 런너 구조 및 특성은 다음과 같다.

1) Open gate structure:

- Controls the nozzle temperature to a constant value
- The structure is simple and the number of constituent parts is small.
- Know how is required for temperature control.
- The gate part can become solidified easily.
- Depending on the plastic, stringiness of the gate plastic can easily occur.

1) 오픈 게이트 구조 :

- 노즐 온도를 일정한 값으로 제어
- 구조가 간단하고 구성 부품의 수가 적음
- 온도 조절을 위한 노-하우가 필요
- 게이트 부분이 쉽게 응고 됨
- 플라스틱에 따라 게이트로부터 늘어지는 실 같은 것이 발생하기 쉽다.

2) On-Off Control structure:

- The gate is heated during injection, and at the end of injection the gate is left to cool
- The structure is relatively simple.
- The temperature control is simple.
- The gate seal is good.
- A special thermocouple is required.

2) 온 - 오프 제어 구조 :

- 사출 중에 게이트가 가열되고, 사출이 끝나면 게이트는 냉각된다.
- 구조가 비교적 간단하다.
- 온도 조절이 간단하다.
- 게이트 씰링이 우수하다.
- 특수한 서머 커플이 필요하다.

3) Hot edge gate structure:

- The gate part is sheared off at the time of opening the mold.
- The structure is relatively simple.
- The gate does not become hard easily.
- There is no stringiness generated.
- There are some restrictions on the applicability of the mold shape.

3) 핫 에지 게이트 구조 :

- 금형을 열 때 게이트 부분이 벗겨진다.
- 구조가 비교적 간단하다.
- 게이트가 쉽게 고화되지 않는다.
- 게이트 실이 생기지 않는다.
- 금형 형상에 따라 적용에 몇 가지 제한이 있다.

4) Valve gate structure:

- The opening and closing of the gate is forcibly controlled by the valve pin.
- The gate seal is definite because it is mechanical.
- It is easy to control the molding conditions.
- A source for driving the opening and closing of the valve pin will be required.

4) 밸브 게이트 구조 :

- 게이트의 개폐는 밸브 핀에 의해 강제 제어된다.
- 게이트 씰링은 기계적이기 때문에 확실하다.
- 성형 조건을 쉽게 제어 할 수 있다.
- 밸브 핀의 개폐를 위한 방안이 필요하다.

- It is necessary to manage the maintenance of the sliding of the valve pin.
- The structure is complex and know how is required even for the mold design.
- The price is high.

- 밸브 핀의 슬라이딩 유지 관리가 필요하다.
- 구조가 복잡하고 금형설계에도 노-하우가 필요하다.
- 가격이 비싸다.

G 026 Determining the Dimensions of the Side Gate

G-026 사이드 게이트의 크기 결정

We describe here the method of determining the dimensions of a side gate that is used very frequently in the molds for injection molding. A side gate is the simplest gate design. Since the molded article and the gate are joined, the gate is cut off in a secondary operation. A side gate can make the molten plastic flow into the cavity in a stable manner, and also since it is possible to apply the dwell pressure sufficiently, it is possible to prevent flashes in the mold.

The side gate dimensions are determined in the following two steps.

여기에는 사출성형을 위한 금형에서 자주 사용되는 사이드 게이트 치수를 결정하는 방법을 설명한다. 사이드 게이트는 가장 간단한 게이트 디자인이다. 성형품과 게이트가 접합되기 때문에 게이트는 2차 가공으로 분리된다. 사이드 게이트는 용융수지를 캐비티 내로 안정적으로 흘려보낼 수 있고, 보합을 충분히 가할 수 있기 때문에 금형의 플래시를 방지 할 수 있다.

사이드 게이트 치수는 다음 두 단계로 결정된다.

First Step: Determining the gate depth

첫 번째 단계 : 게이트 깊이 결정

Table 1

Plastic resin	Coefficient n	Plastic resin	Coefficient n
Polyethylene	0.6	Polycarbonate	0.7
Polystyrene	0.6	PMMA	0.8
Polypropylene	0.7	Polyamide	0.8
Polyacetal	0.7	PVC	0.9

First check the wall thickness t(㎜) of the molded article.
Next, obtain the plastic resin dependent coefficient n from Table 1.

Next, obtain the depth h(㎜) of the side gate using the following equation.

$$h = n \cdot t$$

먼저 성형품의 살 두께 t(㎜)를 확인한다. 다음으로 표 1의 플라스틱 수지 계수 n을 확인한다.

다음 식을 사용하여 사이드 게이트의 깊이 h(㎜)를 구하라.

$$h = n \cdot t$$

Second Step: Determining the gate width

First obtain the surface area A(㎜^{2}) of the molded item.
Next, obtain the molding material dependent coefficient n from Table 1.
Finally, obtain the gate width W(㎜) from the following equation.

$$W = n \cdot \sqrt{A/30}$$

두 번째 단계 : 게이트 너비 결정

먼저 성형품의 표면적 A(㎜^{2})를 구한다. 그 다음에 표 1의 수지 상수 n을 확인한다.

마지막으로 다음 식으로부터 게이트 폭 W(㎜)을 구한다.

$$W = n \cdot \sqrt{A/30}$$

G 027 Determining the Dimensions of the Side Gate (Practice Calculations)

This time we practise the calculations for determining the side gate dimensions explained in the last issue.

G-027 사이드 게이트의 치수 결정(실습 계산)

이번에는 이전에 설명한 사이드 게이트 치수를 계산하기 위한 연습이다.

Question

Let us design the side gate at one location in the case of an injection molded article made of Polyacetal (POM) plastic, when the average wall thickness of the molded article is 1.5 ㎜ and the surface area of the molded article is 4900 $㎜^2$.

In this case, obtain the guidelines for the design dimensions of the gate depth h and width W.

문제

폴리아세탈(POM) 플라스틱으로 성형 된 사출성형품의 경우, 성형품의 평균 살 두께가 1.5 ㎜이고, 성형품의 표면적이 4900 $㎜^2$ 일 때 사이드 게이트를 설계하자.

이 경우, 게이트 깊이 h와 너비 W의 설계 치수에 대한 지침을 구하라.

▶ Sample Answer

The side gate dimensions are determined in the following two steps.

▶ 샘플 답변

사이드 게이트 치수는 다음 두 단계로 결정된다.

First Step: Determining the gate depth

First, we check the wall thickness t(㎜) of the molded item. From the question, we find that t = 1.5(㎜).

Next, we obtain the coefficient n which depends on the molding material from Table 1. In the case of Polyacetal, n = 0.7.

Therefore, the depth h of the side gate will be:

$$h = n \cdot t = 0.7 \times 1.5 = 1.05\ (mm)$$

첫 번째 단계 : 게이트 깊이 결정

먼저, 성형품의 살 두께 t(mm)를 확인한다. 질문에서 t = 1.5(㎜)이다.

다음으로 표 1의 성형 재료에 따라 달라지는 계수 n을 구한다. 폴리아세탈의 경우 n = 0.7 이다.

따라서 사이드 게이트의 깊이 h는 다음과 같다.

$$h = n \cdot t = 0.7 \times 1.5 = 1.05\ (mm)$$

Second Step: Determining the gate width

First we obtain the surface area A($㎜^2$) of the molded item. From the question we find that A = 4900($㎜^2$).

Next, we obtain the coefficient n which depends on the molding material from Table 1. In the case of Polyacetal, n = 0.7.

두 번째 단계 : 게이트 너비 결정

먼저 성형품의 표면적 A($㎜^2$)를 구한다. 질문에서 A = 4900($㎜^2$)이다.

다음으로 표 1의 성형 재료에 따라 달라지는 계수 n을 구하다. 폴리아세탈의 경우 n = 0.7 이다.

Therefore, the width W of the side gate will be:

$$W = n \cdot \sqrt{A/30} = W = 0.7 \cdot \sqrt{4900/30} = 1.63(mm)$$

따라서, 사이드 게이트의 폭(W)은 :

$$W = n \cdot \sqrt{A/30}$$
$$= W = 0.7 \cdot \sqrt{4900/30} = 1.63(mm)$$

In conclusion, considering the ease of machining the mold and its cost, we determine as follows:

$$h = 1(mm) \quad W = 1.6(mm)$$

결론적으로 금형 가공의 용이성과 비용을 고려하여 다음과 같이 결정한다.

h는 1(㎜), W은 1.6(㎜) 이다.

G 028 Varieties of Tunnel Gates

G-028 터널 게이트의 종류

The tunnel gate (submarine gate) is used very frequently as a type of gate having a structure that automatically cuts the molded item and gate at the time of opening and closing the parting surface.

In the basic design of a tunnel gate, although know-how about the shape and size is necessary, we describe the basic varieties concerning the relationship between the molded item and the runners and gates.
The basic patterns of tunnel gates commonly used are shown in the figure.
Classifying broadly into those on the fixed half and those on the moving half of the parting surface, there are four patterns of gate and runner combinations.

터널 게이트(잠수함 게이트)는 파팅면을 개폐할 때 성형품과 게이트를 자동으로 절단하는 구조의 게이트 형태로 많이 사용된다.

터널 게이트의 기본 설계에는 모양과 크기에 대한 노하우가 필요하지만, 성형품과 런너와 게이트 사이의 관계를 고려한 기본적인 종류와 변형된 종류에 대하여 설명한다.

일반적으로 사용되는 터널 게이트의 기본 패턴이 그림에 표시되어 있다.

파팅면을 기준으로 고정측의 터널 게이트와 이동측의 터널 게이트를 크게 분류하면 4가지 종류의 게이트 및 런너 조합이 있다.

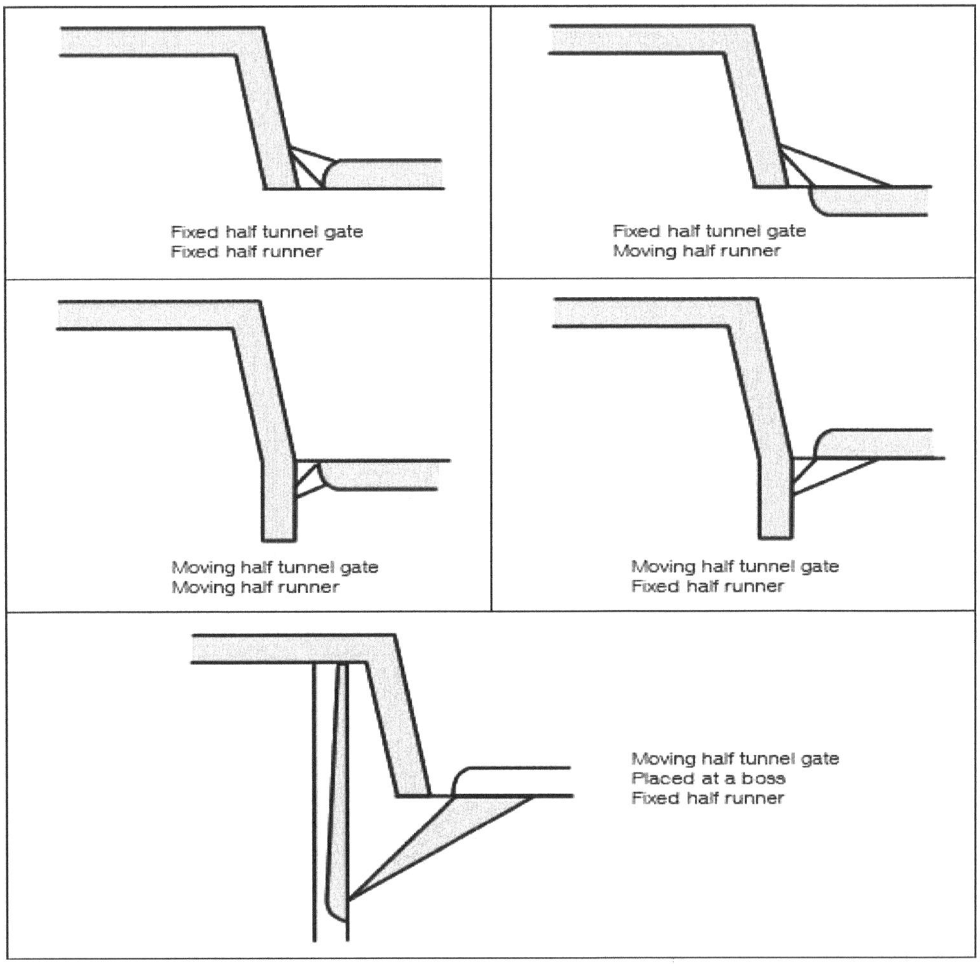

When a tunnel gate is provided on the fixed half, the molded product is cut off from the gate at the time of opening the parting surface. Therefore, the condition of cutting the gate is considered to vary depending on the speed of opening the mold.

On the other hand, when a tunnel gate is provided on the moving half, the molded product is cut off from the gate at the time that the runner ejector pin ejects the runner. Therefore, the condition of cutting the gate is considered to vary depending on the speed of projecting the runner ejector pin.

고정측에 터널 게이트가 설치되어 있는 경우에는 파팅면을 개방할 때 성형품이 게이트로부터 절단된다. 따라서 게이트를 절단하는 조건은 금형을 개방하는 속도에 따라 변하는 것으로 생각된다.

한편, 이동측에 터널 게이트가 설치된 경우, 게이트는 런너 이젝터 핀이 런너를 취출 할 때 성형품에서 절단된다. 그러므로 게이트를 절단하는 조건은 런너 이젝터 핀을 돌출시키는 속도에 따라 변하는 것으로 생각된다.

When the runner is provided on the fixed half, since it is likely that the runner itself remains on the fixed half, a structure that pulls the runner to the fixed side such as a lock pin will be required. When the runner is provided on the moving half, it is necessary to provide an appropriate ejector pin for ejecting the runner.

As a special example, there is also a structure in which a tunnel gate is provided on a boss shape (crafted by scraping the ejector pin) on the moving half, and the plastic is injected from the underside of the top surface of the molded product.

In actuality, the mold design is carried out by considering which gate pattern and runner pattern is appropriate based on the features of the molded product and on the properties of the plastic.

When the specifications require that traces of gate should not remain on the side surface and top surface of the molded product, it will be necessary to provide the gate on the bottom surface of the molded product.

The curved tunnel gate structure is used very rarely as a technique for such situations.

런너가 고정측 금형에 설치될 때, 런너 자체가 고정측에 남아있을 가능성이 있기 때문에 런너 록 핀과 같이 고정측으로 잡아 당기는 구조가 필요하다.

특별한 예로서, 이동측 금형에 보스 형상(이젝터 핀을 긁어서 제작) 터널 게이트를 설치하고, 용융수지는 성형품의 아래쪽으로 부터 주입된다.

실제로, 금형설계는 성형 된 제품의 특징 및 플라스틱의 특성에 기초하여 어떠한 게이트 패턴 및 런너 패턴이 적절한지를 고려하여 수행된다.

사양에 따라 게이트의 흔적이 성형품의 측면과 윗면에 남지 않아야 하므로 성형 제품의 바닥면에 게이트를 설치해야한다. 곡선 터널 게이트 구조는 그런 상황을 위한 기술로 매우 드물게 사용된다.

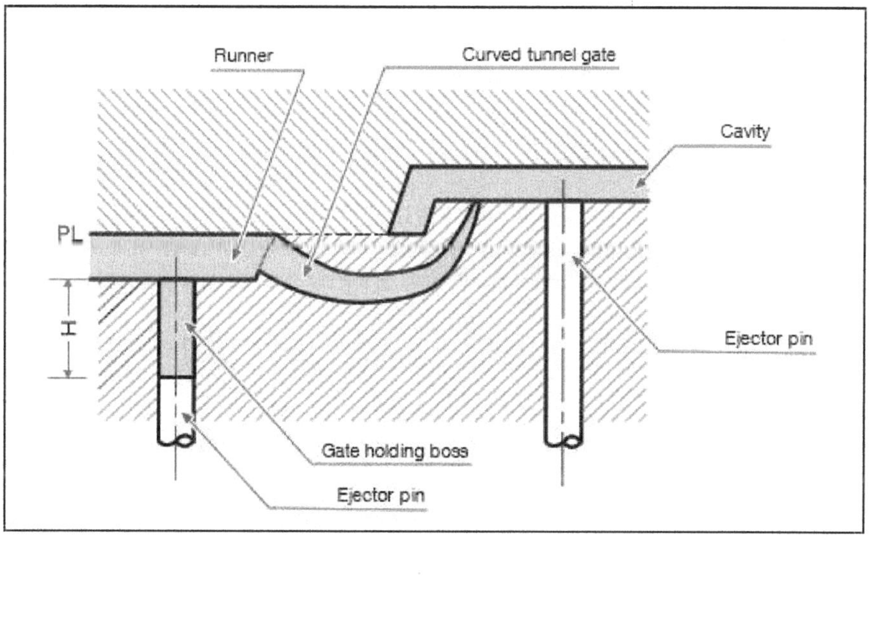

In the curved tunnel gate structure, a gate is provided that has a shape that is curved from the parting line towards the inside of the movable core. Therefore, the gate opening part is positioned at the top surface of the core. An ejector pin is placed at the runner near the gate, and a boss is provided above that pin for supporting the gate. The overall length of the boss, denoted by H, should be selected so that, at the time of ejecting, the gate is supported until the molded product is completely removed from the mold and smooth cutting of the gate is possible.

곡선형 터널 게이트 구조는 이동측 코어의 파팅선으로부터 내부를 향해 만곡된 형상의 게이트가 설치된다. 따라서 게이트 입구는 코어의 상면에 위치한다. 게이트 근처의 런너에 이젝터 핀이 배치되고, 그 핀 위에 보스가 제공되어 게이트를 지지한다. H로 표시된 보스의 전체 길이는 사출시 성형품이 금형에서 완전히 제거 될 때까지 게이트가 지지되어 게이트를 부드럽게 절단 할 수 있도록 선택해야한다.

The shape of the curved part is actually determined by trial and error while making several corrections. However, a wise measure is to use a nested division structure from the beginning so that modifications to the mold can easily be made.

곡선 부분의 모양은 실제로 여러 번의 수정을 하면서 시행착오에 의해 결정된다. 그러나 현명한 방법은 처음부터 중첩 된 분할구조를 사용하여 금형을 쉽게 수정할 수 있도록 하는 것이다.

Since the removal of the gate is affected by the cooling time and the pressure dwell time of the mold, at the time of prototyping it is desirable to vary even these and take samples.

게이트의 제거는 금형의 냉각시간과 보압시간의 영향을 받기 때문에 시험금형에는 이들을 변화시키고 샘플을 취하는 것이 바람직하다.

Whatever method is used, the state of the cut cross-section and the cutting scrap of the tunnel gate can become a problem at the time that the molded product is cut off from the gate.

어떠한 방법이 사용 되더라도 터널 게이트의 절단 단면 및 절단 스크랩의 상태는 성형품이 게이트에서 절단 될 때 문제가 될 수 있다.

The state of the cut cross-section is affected mainly by the following factors.

절단면의 상태는 주로 다음 요소에 의해 영향을 받는다.

1) Design of the shape of the tip part of the gate
2) Thickness of the gate (cross-sectional shape)
3) The distance from the gate cut off part to the runner lock part
4) The state of the effect of dwell pressure
5) The state of polymer orientation at the gate part
6) The timing of cutting off the molded product from the gate

1) 게이트 선단부의 형상 설계
2) 게이트의 두께 (단면 형상)
3) 게이트 절단 부에서 런너 고정부까지의 거리
4) 보압 효과
5) 게이트 부분의 폴리머 배향 상태
6) 게이트로부터 성형품을 절단하는 타이밍

In the case of precision molded products or multiple cavity molds, it is very important to investigate the above factors sufficiently before starting to prepare the molds.

정밀 성형품 또는 복합 성형 금형의 경우 금형 제작을 시작하기 전에 상기 요인을 충분히 조사하는 것이 중요하다.

G 029 Basic Varieties of Tunnel Gates

G-029 터널 게이트의 기본 유형

A tunnel gate (submarine gate) is used very frequently as a gate method for a structure that automatically cuts the product and the gate at the time of opening and closing the parting surface.

터널 게이트 (잠수함 게이트)는 파팅면을 열고 닫을 때 제품과 게이트를 자동으로 절단하는 구조의 게이트 방법으로 매우 자주 사용된다.

While the know how of the shape and dimensions, etc., is necessary at the time of the basic design of a tunnel gate, here, the basic varieties in the relationships among the product, the gate, and the runner are explained.

터널 게이트의 기본 설계 시에 모양과 치수 등이 어떻게 필요한지 알고 있지만, 여기에서는 제품과 게이트, 그리고 런너 사이의 관계를 설명한다.

The basic patterns of tunnel gates generally used are shown in the following figure. When these are broadly classified into a movable side and a fixed side with the parting surface in between them, there are four types of combinations of the gate and runner.

일반적으로 사용되는 터널 게이트의 기본 패턴은 다음 그림과 같다. 이들을 파팅면을 사이에 두고 가동측과 고정측으로 크게 구분하면 게이트와 런너의 조합은 4 종류가 있다.

When the tunnel gate is provided on the fixed side, cutting of the molded product and the gate is done at the time when the parting surface is opened. Therefore, the condition of gate cutting is considered to vary depending on the speed of opening the mold.

터널 게이트가 고정측에 설치 된 경우, 성형품 및 게이트의 절단은 파팅면이 개방 될 때 이루어진다. 따라서 게이트 절단 조건은 금형이 열리는 속도에 따라 변하는 것으로 생각된다.

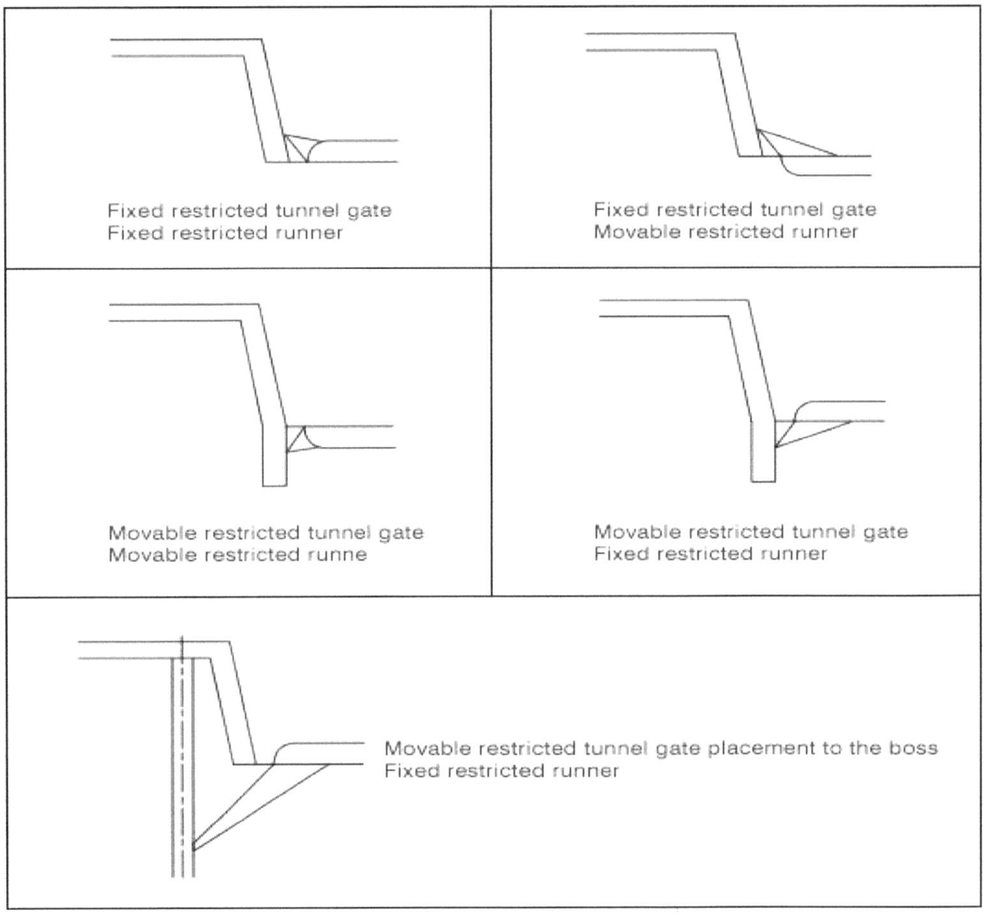

On the other hand, when the tunnel gate is provided on the movable side, the cutting of the molded product and gate is done at the time when the runner ejector pin protrudes from the runner. Therefore, the condition of gate cutting is considered to vary depending on the speed of the protruding runner ejector pin.

반면에, 터널 게이트가 가동측에 설치되면, 성형품 및 게이트의 절단은 런너 이젝터 핀이 취출 될 때 이루어진다. 따라서 게이트 절단 조건은 런너 이젝터 핀의 취출 속도에 따라 달라진다.

When the runner is provided on the fixed side, since it is likely that the runner itself remains on the fixed side, it is necessary to have a structure with a long pin in which case the runner is pulled to the movable side.

런너가 고정측에 있는 경우, 런너 자체가 고정측에 남아있을 가능성이 있기 때문에 런너가 가동측으로 당겨지는 긴 핀이 필요하다.

When the runner is provided on the movable side, it is necessary to provide an appropriate ejector pin for protruding the runner.

런너가 가동측에 있는 경우 런너를 취출 시키기 위한 적절한 이젝터 핀이 필요하다.

As a special example, there is a structure in which the tunnel gate is provided on a boss shape (formed by fine grinding the ejector pin) located on the movable side, and the plastic is injected from the underside of the top surface of the molded product.

In actuality, depending on the features of the molded product or the characteristics of the plastic, the design of the mold is carried out considering which gate pattern and which runner pattern is most appropriate.

특별한 예로, 터널 게이트가 가동측에 위치하는 보스 형상(이젝터 핀을 정밀하게 연삭된 이젝트 핀)으로 설치되면, 주입된 플라스틱 수지가 이젝트 핀을 따라 상측으로 주입되는 구조가 있다.

실제로, 성형품의 특성이나 플라스틱의 특성에 따라서 금형을 설계할 때에는 가장 적합한 게이트와 런너 패턴을 고려하여야 한다.

G 030 Designing the Tip Shape of Pin Point Gate

G-030 핀 포인트 게이트의 선단 형상 설계

The problems that can occur in the case of pin point gate structure are the following.

1) he tip of the gate remains as a projection on the surface of the molded product, or scratches out a part of the molded product.
2) The filling does not progress smoothly in spite of the filling pressure or the dwell pressure being high.

These are the problems that a mold designer has to solve whenever adopting a pin point gate.
Some technical measures for solving these problems are described below.

Fig. 1 shows the structure of a general pin point gate. This is the shape when the gate design is made without any special considerations.

핀 포인트 게이트 구조의 경우 발생할 수 있는 문제점은 다음과 같다.

1) 게이트의 선단은 성형품의 표면에 돌기로서 남아 있거나 성형품의 일부분을 긁어낸다.
2) 충전압력 또는 보압이 높더라도 충전이 원활하게 진행되지 않는다.

이들은 금형설계자가 핀 포인트 게이트를 채택 할 때마다 해결해야 하는 문제이다. 이러한 문제를 해결하기 위한 기술적인 조치는 아래에 설명되어 있다.

그림 1은 일반적인 핀 포인트 게이트의 구조를 보여준다. 이는 특별히 고려하지 않고 게이트 설계를 수행 할 때의 형상이다.

[Fig. 1]

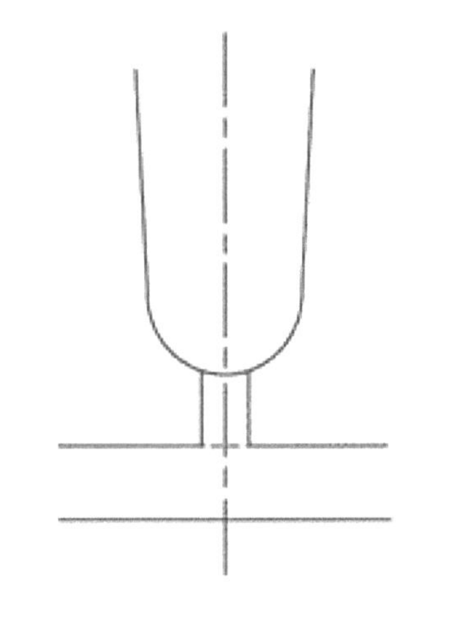

[Fig. 2]

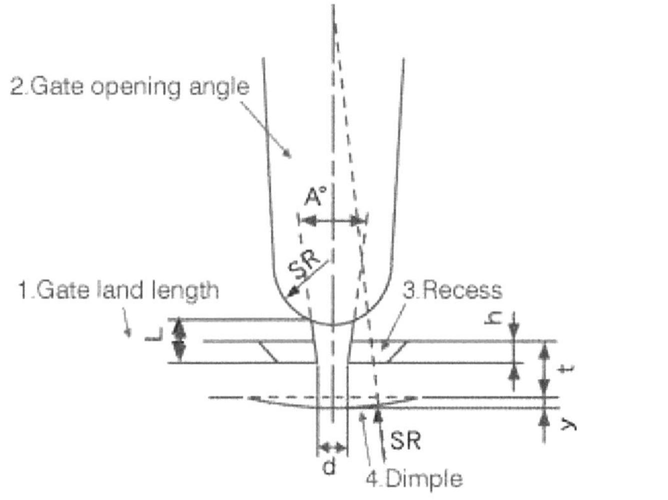

On the other hand, Fig. 2 is a design in which measures have been taken to solve the above problems.

반면에 그림 2는 위의 문제를 해결하기 위한 방안을 고려한 디자인이다.

Key factor 1: Gate land length L

핵심 요소 1 : 게이트 랜드 길이 L

If the gate land length is longer than necessary, at the time of cutting the gate part, it is possible that the gate gets cut in the middle thereby leaving a projection on the molded product. As a rule of experience, it is reco mmended that the gate land length L is in the range of about 1 to 2 times the gate tip diameter d.

게이트 랜드 길이가 필요한 것보다 길면 게이트 부분을 절단 할 때 게이트가 중간에서 절단되어 성형품에 돌출부가 남아있을 수 있다. 경험상 게이트 랜드 길이 (L)는 게이트 팁 직경 (d)의 약 1 ~ 2배 범위이다.

Key factor 2: Gate opening angle A

핵심 요소 2 : 게이트 개방 각도 A

As an opening angle, a gate is provided with a conical shaped taper. If it has a conical shape, at the time of cutting the gate, the possibility becomes high that the gate is always cut at the minimum cross-section part where the gate and the molded product meet. In

개방 각도로서 게이트는 원추형 테이퍼를 갖는다. 원추형인 경우에는 게이트를 절단 할 때 게이트와 성형품이 만나는 최소 단면 부분에서 항상 게이트가 절단 될 가능성

addition, even releasing from the mold becomes easier. Generally, the value of A is about 15 to 30° . Although the cutting becomes definite when the value is large, there is also a trend that the wear of the tip part progresses faster.

이 높아진다. 또한 금형에서 떨어지기도 더 쉽다. 일반적으로 A의 값은 약 15 ~ 30° 이다. 값이 클 때 절단면이 명확해 지지만 팁 부분의 마모가 빠르게 진행되는 경향이 있다.

Key factor 3: Recess

If the gate is made to penetrate more into the molded product by forming a recess at the surface where the gate meets the molded product, even if there is a projection at the cut part, it will not project beyond the surface of the molded product.

For providing a recess, it is always necessary to get the permission in the drawing for the change in the specifications from the designer of the molded product.

핵심 요인 3 : 오목 홈

게이트가 성형품과 만나는 곳에서 오목한 홈으로 인하여 게이트가 성형품 안쪽으로 만들어지면, 절단 부분에 돌기가 있어도 성형품의 표면을 넘어서 돌출하지 않는다.

오목 홈을 형성시키기 위해서는 항상 제품 설계자로부터 사양 변경에 대한 도면 허락을 받아야 한다.

Key factor 4: Dimple

A dimple is a spherical shaped depression provided on the side opposite the gate of the same extent as the ordinary wall thickness of the molded product so that the molten plastic can flow in a stable manner.

Even for providing a dimple, it is always necessary to get the permission in the drawing for the change in the specifications from the designer of the molded product.

In the above manner, if considerations are given in the design aspects, and also, if the gate has been machined accurately using electric discharge machining, etc., there is a very high probability that the above problems are solved.

핵심 요소 4 : 딤플

딤플은 성형품의 평균적인 살 두께와 같이 제이트의 반대쪽에 만들어지는 구형상의 오목부로서 이것은 용융 수지가 안정적으로 흐를 수 있게 한다.

딤플을 설계하는 경우에도 제품 설계자로부터 사양변경에 대한 도면 허락을 받아야 한다.

이와 같이 설계측면에서 고려되고, 또한 방전 가공 등을 이용하여 정확하게 게이트를 가공하면 상기 문제가 해결 될 가능성이 매우 높다.

G 031 Fool Proof in Mold Design

Literally, the meaning of "fool proof" has the nuance of "prevention of fooling". In concrete terms, this refers to a construction or shape that prevents wrong assembly or disassembly by humans making an inadvertent mistake.

Taking the example of a mold, consider that there are two core pins A and B that are extremely alike. The position of incorporating in the main core is decided.

If, the mold was disassembled and maintenance work was done on it, there is naturally the possibility that the assembling positions A and B are reversed while assembling again the mold. It is not possible to eliminate completely inadvertent mistakes or misunderstanding even in the case of highly experienced persons.

In view of this, if the flange parts of the core pins are cut in different shapes and the assembly holes of the main core are also machined to those shapes, since it will only be possible to assemble the core pins A and B only at the correct positions irrespective of who does the assembling, there can be no wrong assembly. These kinds of measures are called "fool proof" measures.

When there are parts that are likely to be assembled wrongly, or when there is a sequence for assembling, it is necessary to make fool proof designs so that there is no inadvertent mistake by the operator while assembling.

G-031 풀 프루프 금형설계

문자 그대로 "fool proof"의 의미는 "어리석은 행위 예방"이라는 뜻이다. 구체적으로는 사람이 의도치 않게 잘못 조립하거나 잘못 분해하는 것을 방지하는 구조 또는 모양을 의미한다.

금형을 예로 들면, 두 개의 아주 비슷한 코어 핀 A와 B가 있다. 메인 코어에 결합 위치가 결정된다.

금형을 분해하여 유지 보수 작업을 한 경우, 금형을 다시 조립하는 동안 A와 B의 조립 위치가 뒤바뀔 가능성이 자연스럽게 존재한다. 고도로 숙련 된 사람이라 할지라도 우발적인 실수나 잘못된 이해를 완전히 배제 할 수는 없다.

그래서, 코어 핀의 플랜지부를 다른 형상으로 절단하고, 또한 메인 코어의 조립 구멍을 그 형상으로 가공하면 코어 핀 (A, B)만이 조립 가능하기 때문에 어느 누가 조립하는지와 관계없이 정확한 위치, 잘못된 조립이 있을 수 없다. 이러한 종류의 조치를 "fool proof"라고 한다.

잘못 조립 될 가능성이 있는 부품이 있거나, 조립 순서가 정해져 있는 경우, 조립하는 동안 작업자의 부주의 한 실수를 하지 않도록 풀 프루프 설계가 이루어 져야한다.

If a mistake is made in assembling a large sized mold or a mold for export, carrying out that work again can take a long time. In addition, by clamping a wrongly assembled mold can break the mold which can lead to a huge loss of money, etc.

For judging whether or not fool proof measures are required, it is very important to give training to mold designers not only about simple cost reduction considerations but also about making reviews regarding fool proof measures.

대형 금형이나 수출용 금형을 조립하는 데 실수가 있으면 다시 작업하는 데 오랜 시간이 걸릴 수 있다. 또한 잘못 조립 된 금형을 클램핑하면 금형이 손상되어 막대한 금전 손실 등을 초래할 수 있다.

풀 프루프 대책이 필요한지 아닌지 판단하기 위해서 단순한 비용 절감을 고려할 뿐만 아니라 풀 프루프 대책에 대한 검토를 하는 것이 중요하다.

G 032 Method of Providing Air Vents

G-032 에어벤트를 설치하는 방법

Air is present in the space in the cavity of a mold. In injection molding, molten plastic is poured inside the cavity, that is, it is also possible to say that the process is that of replacing air with molten plastic.

Therefore, if it is possible to discharge the air from inside the cavity to the outside efficiently, then it is possible to make small the resistance at the time of filling; therefore it is possible to suppress the filling pressure to a low value, and as a result, it is possible to make small the residual stress in the molded product. In addition, it is also possible to prevent the breakage of core pins, etc., due to the filling pressure.

On the other hand, if it is possible to discharge from inside the cavity to the outside the volatile constituents or gas generated from molten plastic, it is possible to improve the quality of the molded product.

공기가 금형의 캐비티 내에 존재한다. 사출성형에서 용융 된 플라스틱이 캐비티 내부에 사출된다. 즉, 공기를 용융수지로 대체하는 공정이라고 말할 수 있다.

따라서, 캐비티 내부의 공기를 효율적으로 외부로 배출 할 수 있으면 충전시의 저항을 작게 할 수 있고, 충전압력을 낮게 억제 할 수 있고, 결과적으로 성형품의 잔류 응력을 작게 할 수 있다. 또한, 충전 압력에 의한 코어 핀의 파손 등을 방지 할 수 있다.

한편, 캐비티의 용융수지로부터 발생하는 휘발성 성분이나 가스를 캐비티 내부로부터 외부로 배출 할 수 있다면, 성형품의 품질을 향상시킬 수 있다.

"Air vent" is one that carries out these functions. An air vent can be provided on the parting surface as shown in the example in Fig. 1, or can be provided on the core pin side surface as shown in the example in Fig. 2.

"에어벤트"는 이러한 기능을 수행하는 것이다. 에어벤트는 그림 1의 예와 같이 파팅면이 될 수도 있다. 그림 2와 같이 코어 핀 측 표면에 설치해도 된다.

[Fig. 1] Example of an air vent provided on the parting surface

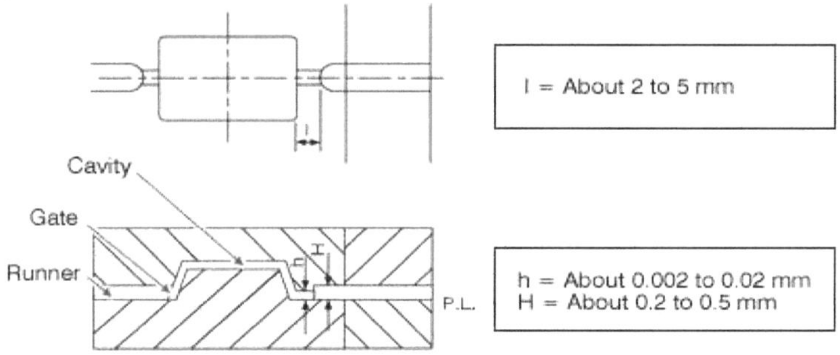

[Fig. 2] Example of an air vent provided on the core pin

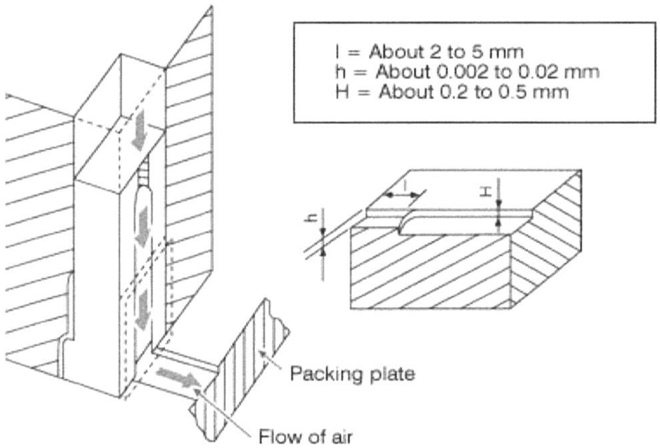

One end part of the air vent should be in contact with the cavity. The depth of the air vent at this contacting part is generally about 0.002 to 0.02 ㎜. Although the efficiency of discharging air or gas becomes better when this is deep, care should be taken because it is possible that burrs are generated in the molded product. Select an appropriate value depending on the plastic material and the cavity surface temperature, etc.

에어 밴트의 한 쪽 끝이 캐비티와 접촉해야 한다. 접촉부에서의 에어벤트 깊이는 일반적으로 약 0.002 ~ 0.02 ㎜이다. 이것이 깊어지면 공기나 가스의 배출 효율이 좋아 지지만, 성형품에 버가 생길 가능성이 있으므로 주의가 필요하다. 플라스틱 재질 및 캐비티 표면 온도 등에 따라 적절한 값을 선택한다.

It is sufficient if the length of the air vent at this part is about 2 to 5 ㎜.

The important factor is that the other end part of the air vent is connected to a much deeper air vent. The depth of this deeper air vent should be about 0.2 to 0.5 ㎜ and becomes a path for the air or gas to be exhausted to atmosphere from the side surface of the mold plate or of the packing plate.

Since slag or tar components generated from the plastic during injection molding is deposited in the air vent, it is necessary to clean the air vents periodically, such as about once in a fortnight to once in a month.

The position of providing the air vent can be near a part where a weld is generated, near the final last filling part, or in the middle of the runner. It is also a practical approach to make one trial and then decide the detailed location of providing the air vents.

이 부분에서 에어벤트의 길이는 약 2 ~ 5 ㎜이면 충분하다.

중요한 요소는 에어벤트의 다른 끝 부분이 더 깊은 에어벤트 출구에 연결된다는 것이다. 더 깊은 에어벤트의 깊이는 약 0.2 ~ 0.5 ㎜가 되어야하며, 공기 또는 가스가 형판 또는 패킹 플레이트의 측면에서 대기로 배출되는 경로가 된다.

사출성형 할 때 플라스틱으로부터 생성된 슬래그 또는 타르 성분은 에어벤트에 침전되므로 정기적으로 에어벤트를 청소해야 한다.(예 : 2주에 1회 정도, 한 달에 1회 정도).

에어벤트를 설치하는 위치는 웰드라인이 생성되는 부분, 최종 충전부 근처 또는 런너의 중간 부분 근처일 수 있다. 또한 한 번의 시험사출을 거치고, 에어벤트의 위치를 결정하는 것도 실용적인 접근법이다.

G 033 Hints on Mold Design for Plastics with High Cavity Surface Temperatures

G-033 고온 캐비티 표면이 필요한 금형의 설계 참고사항

There are more and more engineering plastics and super engineering plastics for which the cavity surface temperatures exceeds 100℃.

When the surface temperature of the cavity exceeds around 90℃, normally, raising the temperature becomes difficult when the temperature control is done using water (hot water).

Generally, the following methods are used.

캐비티 표면 온도가 100℃를 초과하는 엔지니어링 플라스틱 및 수퍼 엔지니어링 플라스틱이 점점 더 많이 등장하고 있다.

캐비티 표면 온도가 약 90℃를 초과하면 일반적으로 물(온수)을 사용하여 온도 제어를 하면 온도 상승이 어렵다.

일반적으로 다음과 같은 방법이 사용된다.

1. Oil temperature control

In temperature control using oil, the temperature is maintained constant by re-circulating oil from a circulating pump by passing it through joints and hoses and a flow path provided in the mold plate or cavity. Once the temperature rises up to the preset temperature, it is possible to maintain a relatively stable temperature.

However, there is the drawback that raising the temperature takes time. In addition, there is the danger of burns in handling the oil, and also the after-treatment of oil is cumbersome.

1. 오일 온도 제어

오일을 이용한 온도 제어에는 순환 펌프의 오일을 조인트와 호스 및 형판 또는 캐비티에 제작된 유로를 통과시키고 재순환하여 온도를 일정하게 유지한다. 일단 온도가 미리 설정된 온도까지 상승하면 비교적 안정된 온도를 유지할 수 있다.

그러나 온도 상승에는 시간이 걸리는 단점이 있다. 또한, 오일 운반 시 화상의 위험이 있고, 또한 오일의 후 처리가 복잡하다.

2. Electric heater control

In temperature control using an electric heater (cartridge heater), the temperature is maintained constant by using the heater in combination with a temperature sensor (thermo couple, etc.). Since the thermal capacity is high, there is the advantage that raising the temperature is fast.

However, the temperature in the vicinity of the heater is high and the temperature becomes low at locations distant from the heater, and it is difficult to maintain a uniform temperature distribution.

In addition, the life of the heater is limited and it needs to be replaced periodically.

The mounting holes and the clearance are very important in installing a heater. If the clearance is large, the heater gets heated excessively and hence its life gets reduced. The Misumi M-HTM3021 (Mold temperature controller for cartridge heater) can control the heater temperature using the PID control method, and extremely stable temperature control becomes possible more than an ON-OFF control type controller.

2. 전기 히터 제어

전기 히터(카트리지 히터)를 사용하는 온도 제어에는 온도 센서 (써머 커플 등)와 함께 히터를 사용하여 온도를 일정하게 유지한다. 열용량이 크기 때문에 온도 상승이 빠르다는 장점이 있다.

그러나 히터 부근의 온도가 높고, 히터로부터 떨어진 위치에서는 온도가 낮아져 균일한 온도 분포를 유지하는 것이 어렵다.

또한 히터의 수명이 제한되어 주기적으로 교체해야 한다.

설치 구멍 및 간격은 히터 설치 시 매우 중요하다. 공차가 크면 히터가 과도하게 가열되어 수명이 단축된다. 미스미 M-HTM 3021(카트리지 히터용 금형 온도 컨트롤러)는 PID 제어 방식으로 히터 온도를 제어 할 수 있으며, ON-OFF 제어 방식의 컨트롤러보다 매우 안정된 온도 제어가 가능하다.

In order to carry out temperature control more accurately, it is recommended to install a thermal insulating plate between the platen of the injection molding machine and the installation plate of the mold. Or else, it is also possible to install thermal insulating plates in the periphery of the mold plate.

보다 정확하게 온도 제어를 하기 위해서 사출성형기의 설치판과 금형의 설치판 사이에 단열판을 설치하는 것이 좋다. 또는 형판 주변에 단열판을 설치할 수도 있다.

Name of plastic	Cavity surface temperature (℃)
PPS with 30% glass fiber (Polyphenylene sulfide)	130 to 150
LCP with 40% glass (Liquid crystal polymer)	70 to 110
PET (Polyethylene terephthalate)	130 to 150
PA46 (Polyimide / Nylon-46)	80 to 120
PC (Polycarbonate)	80 to 120

G 034 Treatment of Undercuts using Insert Blocks

G-034 인서트 블록을 이용한 언더컷 처리

When the product has an undercut, it is removed from the mold using mechanisms such as a sliding core, or a slant projection, or forced projection, etc.

However, in the case of complex undercut shapes, it can also be very difficult to solve the problem using mechanical structures.

In such situations, it may be possible to solve the problem using a traditional method called "insert block structure".

제품에 언더컷이 있는 경우 슬라이딩 코어, 경사 밀핀 또는 강제 빼기 등과 같은 메커니즘을 사용하여 금형에서 제거된다.

그러나 복잡한 언더컷 형상의 경우 기계 구조를 사용하여 문제를 해결하는 것이 매우 어려울 수 있다.

이러한 상황에는 "블록 구조 삽입"이라는 전통적인 방법을 사용하여 문제를 해결할 수 있다.

An "insert block structure" is one in which the core is divided at the undercut part, the divided core(insert block) is taken out along with pushing out the molded product, and after that, the insert block is removed by hand.

At the time of the next molding, the insert block is set inside the mold again.

Therefore, in the case of an insert block structure there is the drawback that the molded product becomes a semi-automatically prepared item.

However, since it is possible to provide the desired undercut shape in the integrally molded product, this is a very useful method in special cases when such special structures are required.

In summary, the molded product in which the advantages of the insert block structure are brought out are the following.

1) Molded product in which it is not possible to handle undercuts using the ordinary slide cores, etc.
2) Molded product having undercut shapes, but whose production volume is small.
3) Sample molded product

Since an insert block is installed and removed manually, techniques such as machining or chamfering the escape for guiding the block inserting part, and the mating tolerances will be the know how of the manufacturer.

It is not always necessary that molds must have structures suitable for automatic production, and for designers in the digital age it is very important to know that it is possible to obtain the desired molded product by incorporating manual steps that are more analog than digital.

「인서트 블록 구조」란 코어가 언더컷 부분에서 분할되고, 성형품을 밀어 내면서 분할 코어(인서트 블록)를 꺼내고, 그 후 인서트 블록을 손으로 떼어내는 것이다.

다음 성형 할 때에 삽입 블록은 다시 금형 내부에 설치된다.

따라서, 인서트 블록 구조의 경우, 성형 된 제품이 반자동으로 준비되는 단점이 있다.

그러나 완성된 성형품에 원하는 언더컷 형상을 부여할 수 있으므로 특별한 구조가 요구되는 특수한 경우에 대하여 매우 유용한 방법이다.

요약하면, 인서트 블록 구조의 장점을 발휘하는 성형품은 다음과 같다.

1) 일반 슬라이드 코어 등으로 언더컷을 처리 할 수 없는 성형 제품
2) 언더컷 모양이지만 생산량이 적은 성형 제품.
3) 샘플 성형품

인서트 블록은 수동으로 설치 및 제거되므로 블록 삽입부를 안내하기 위한 탈출구 가공, 모따기 및 결합 공차들은 제조업체의 노하우가 된다.

금형이 자동 생산에 적합한 구조를 가져야하는 것은 항상 필요하며, 디지털 시대의 설계자에게는 디지털보다 더 아날로그인 수동 단계를 통합하여 원하는 성형 제품을 얻을 수 있다는 사실을 인식해야 한다.

G 035 Ejecting Stripper Plate Structure

The ejecting stripper structure is a structure that is used when there should be no traces of the ejector pin on the molded item. This is used for transparent optical components made of acrylic or polycarbonate, etc., such as that shown in Fig. 1.

G-035 스트리퍼 플레이트 구조의 취출

스트리퍼 구조의 취출은 성형품에 이젝터 핀의 흔적이 없어야 할 때 사용되는 구조이다. 이는 그림 1과 같은 아크릴 또는 폴리카보네이트 등으로 만들어진 투명한 광학 제품에 사용된다.

[Fig.1] Example of molded item suitable for ejecting stripper

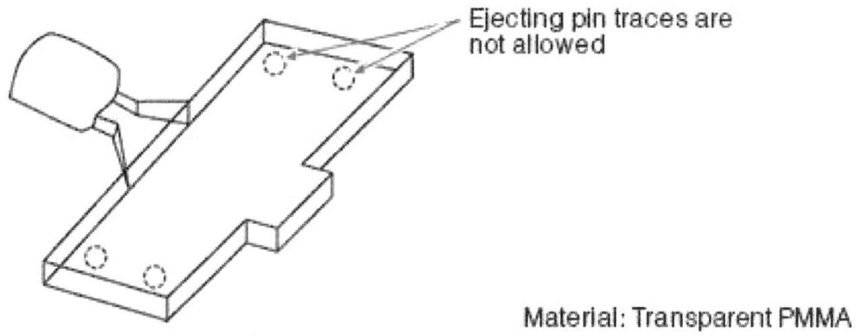

[Fig.2] Example of the structure of a stripper plate mold

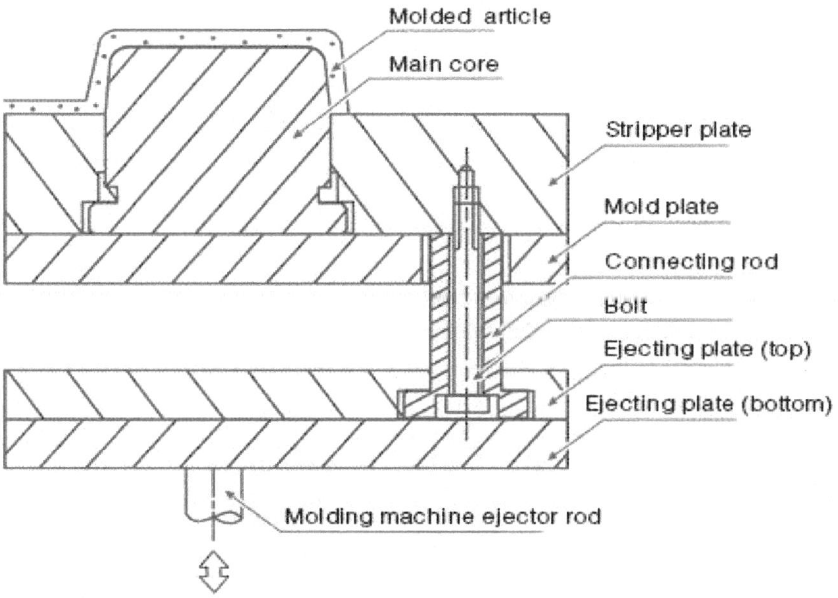

Fig. 2 shows an actual example of the ejecting stripper structure

Firstly, the main core that forms the molded article is fixed to the moving half mold plate using bolts, etc.
The stripper plate is placed so that it envelopes the side surface of the main core, the stripper plate is further connected to the ejector plate via a connecting rod.
In other words, when the ejector plate moves forward, even the connecting rod moves forward along with it, thereby causing the stripper plate to move forward and ejecting the molded article.

As the ejector plate returns, the stripper plate also returns to the original position.

그림 2는 실제 스트리퍼 취출의 예를 나타낸다.

우선, 성형품을 성형하는 메인 코어를 볼트 등을 사용하여 이동측 금형 형판에 고정한다. 스트리퍼 플레이트는 메인 코어의 측면을 둘러 싸도록 배치되고, 스트리퍼 플레이트는 이젝터 플레이트에 연결된다. 다시 말하면, 이젝터 플레이트가 전진 할 때, 커넥팅로드도 함께 이동하여 스트리퍼 플레이트가 전방으로 이동하여 성형품을 취출한다.

이젝터 플레이트가 돌아오면 스트리퍼 플레이트도 원래 위치로 돌아간다.

[Fig.3] Diagram showing the state of operation of the mold

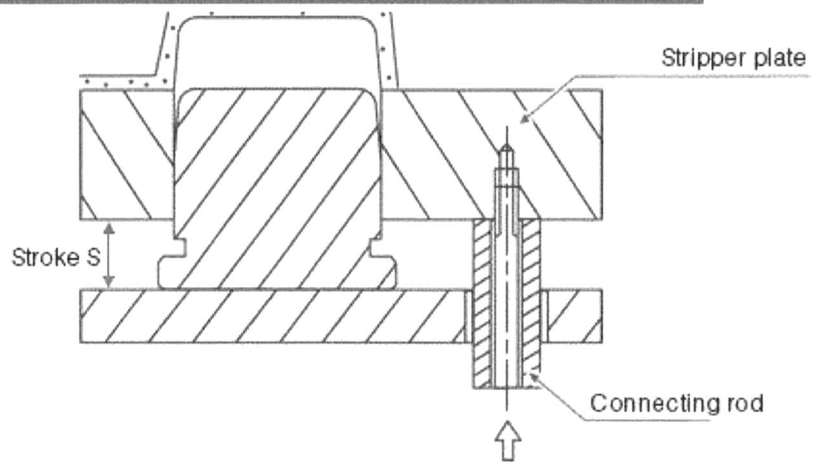

At the time of carrying out the basic design of the stripper plate structure, while it is of course necessary to prepare properly the drawing of the operating states, in addition it is very important for smooth operation to take measures such as managing the clearance between the stripper plate and the main core, tapered escape, etc.

스트리퍼 플레이트 구조의 기본 설계를 수행 할 때, 물론 작동 상태의 도면을 적절하게 준비 할 필요가 있지만, 또한 스프리퍼 플레이트와 메인 코어 사이의 클리어런스 관리, 테이퍼형 탈출과 같은 방안을 원활하게 진행하는 것이 매우 중요하다.

In addition, adoption of a guide system, selection of the steel material for the stripper plate, heat treatment, etc., also become very important factors.

또한, 가이드 시스템의 선택, 스트리퍼 플레이트용 강재 선택, 열처리 등은 매우 중요한 요소이다.

[Fig.4] Precautions about the ejecting stripper structure

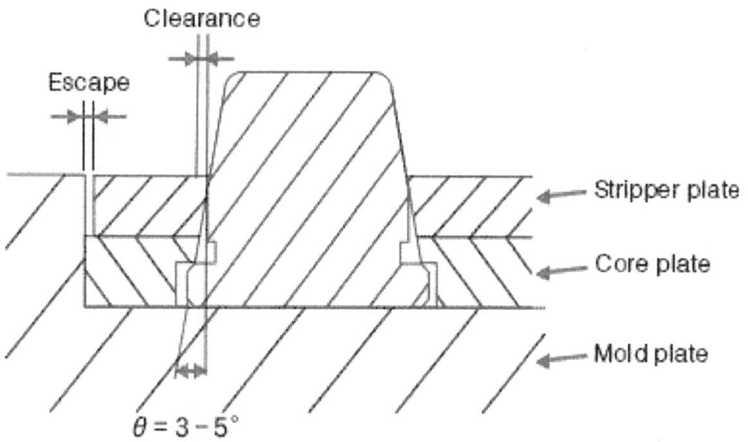

G 036 Gate Cutting Unit inside the Mold

G-036 금형 내부의 게이트 절단 장치

When injection molding is done with a cold runner, this is a unit product for cutting off the gate using the operation of opening the mold.

Side gates or submarine gates having large gate diameters are used for the gates of thin walled molded article or of molded article in which shrink is not acceptable. But when this is done, not many people would have found it troublesome to carry out cutting of the gates in the subsequent processes.

콜드 런너로 사출성형을 할 때 금형이 열리는 것을 이용하여 게이트를 절단하는 장치이다.

큰 치수의 사이드 게이트와 서브머린 게이트는 얇은 살 두께의 성형품 또는 수축이 허락되지 않는 성형품의 게이트로 사용된다. 그러나 이 게이트의 사용 후에는 후속 작업에서 게이트를 절단해야 하는 번거로움이 있다는 것을 아는 사람은 많지 않다.

The "Gate cutting unit inside the mold" is a mechanism by which the gate part is rotated along a spiral groove with the parting surface still closed thereby cutting the gate.

Since the cutting can be done inside the mold, the gate cut surface becomes uniform, and also, it is possible to suppress the generation of gate shreds at the time of cutting.

The following types of structures can be used for the gate.

1) Side gate
2) Submarine gate
3) Jump gate (overlap gate)

In particular, its effect can be expected during the formation of caps, lids, and cases of containers. The effect is further enhanced when the volume of production is medium to large.

The diameter of the rotating gate pin comes in two types, φ16 ㎜ and φ20 ㎜. High speed steel is used as the material for this.

The materials for which this mechanism can be used are - ABS, PS, PA6, PA66, PBT, POM, PPS, PC, etc.
(This is not suitable for acrylic resin etc. in which case cracks are generated during cutting.)

Since there are limitations on the size of the mold, plate thickness, etc., at the time of considering the use of this mechanism.

When it is difficult to use a hot runner gate due to limitations on the gate position, etc., or when a side gate, etc., must be provided this offers a new option.

"금형 내부의 게이트 절단 장치"는 파팅면에 여전히 닫혀있는 스파이럴 그루브를 따라 회전하면서 절단하는 메카니즘이다.

또한, 금형 내에서 절단 할 수 있기 때문에 게이트 절단면이 균일해지고, 또한 절단 시에 게이트 갈라짐의 발생을 억제 할 수 있다.

다음 유형의 구조를 게이트에 사용할 수 있다.

1) 사이드 게이트
2) 서브머리 게이트
3) 점프 게이트 (오버랩 게이트)

특히, 캡, 뚜껑 및 용기의 경우에 그 효과가 기대된다. 생산량이 중대형 일 때 효과가 더욱 향상된다.

회전 게이트 핀의 직경은 ∅16 ㎜와 ∅20의 두 가지 유형이 있다. 이를 위해 고속도강이 사용된다.

이 메커니즘을 위한 재료는 ABS, PS, PA6, PA66, PBT, POM, PPS, PC 등이다. (절단 중에 균열이 발생하는 아크릴 수지 등에는 적합하지 않다.)

이 장치의 사용을 고려할 때 금형, 판 두께 등의 크기에는 제한이 있다.

게이트 위치 등의 제한으로 인해 핫 런너 게이트를 사용하기가 어려울 때 또는 사이드 게이트 등을 제공해야하는 경우 이는 새로운 옵션을 제공한다.

The cost of cutting the gate becomes quite a substantial amount when the volume of production increases. This product brings out its advantages when trying to reduce these costs.

게이트를 절단하는 비용은 생산량이 증가 할 때 상당한 금액이 된다. 이 게이트 절단 장치는 이러한 비용을 줄이는 장점을 가졌다.

G 037 Cavity and Core Integrated Structure and Spliting Structure

G-037 캐비티 및 코어의 통합구조 및 분할구조

The cavity and the core are very important parts in forming the shape of the molded product, and the method of their preparation can be divided broadly into two types. These types of the method of preparing are "Integrated Structure" and "Spliting Structure".

The "integrated structure" is one in which, as the name implies, the cavity and the core are prepared as a single integrated component.

The "Spliting structure" is one in which the cavity and the core are divided into 2 or more components which are assembled together.

Each of these methods has its own advantages and disadvantages which are listed in Table 1. Further, some actual examples of the different structures are shown in Fig. 1

캐비티 및 코어는 성형품의 형상을 형성하는 데 있어서 매우 중요한 부분이며, 그 제작 방법은 크게 두 가지 유형으로 나뉠 수 있다. 이러한 준비 방법은 "통합구조"와 "분할구조"이다.

"통합구조"란 이름에서 알 수 있듯이 캐비티와 코어가 단일 요소로 준비되어 있는 구조이다.

"분할구조"는 캐비티와 코어가 함께 조립 된 2 개 이상의 구성 요소로 분할 된 구조이다.

각각의 방법이 가진 장단점을 표 1에 나타내었다. 또한, 상이한 구조들의 실제 예가 그림 5에 나타내었다.

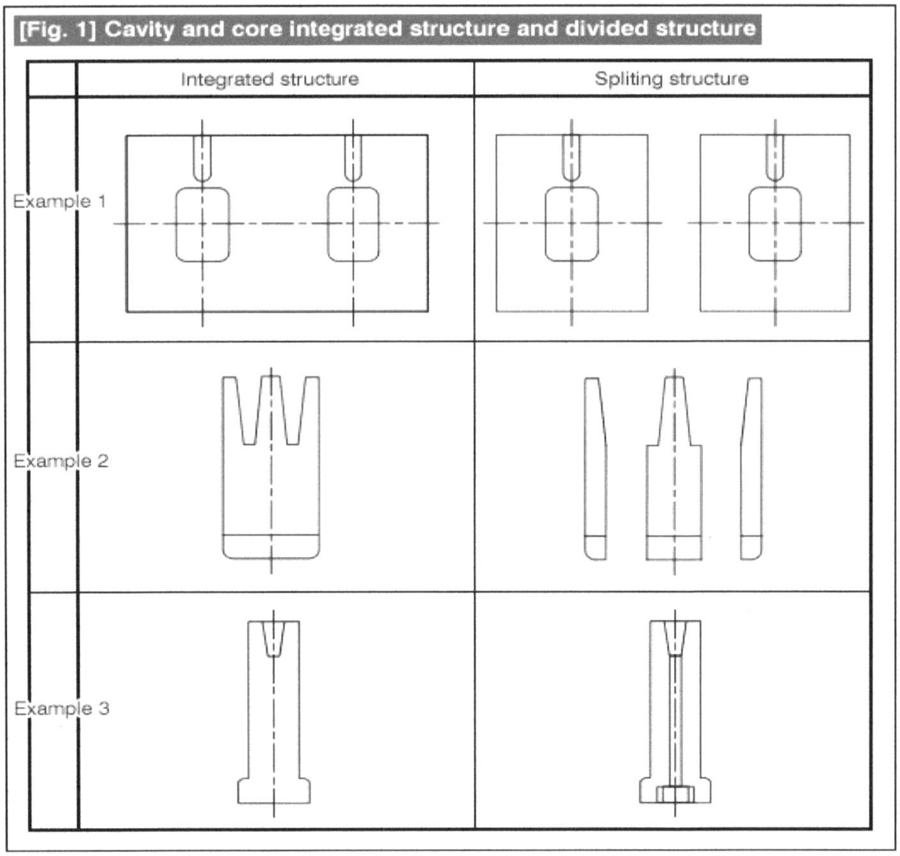

Fig. 1 Cavity and core integrated structure and divided structure

[그림 1] 캐비티와 코어의 통합구조 및 분할구조

(Evaluation: ◎ = Excellent, ○ = Good, △ = Partially satisfactory, ▲ = Not suitable.

(평가 :. ◎ = 우수, ○ = 좋음, △ = 부분적으로 만족, ▲ = 적당하지 않음)

Comparison item	Integrated structure	Divided structure
Component count	◎	△
Ease of machining	◎ – △	◎ – △
Cost of machining	◎ – △	◎ – △
Material cost	○	△
Corner part shape accuracy	△	◎
Ease of polishing	△	◎
Air vent effect	△	◎
Ease of maintenance	△	◎

비교 항목통합구조분할구조부품 수가공의 용이성

The decision of whether to use an integrated structure or a Split structure should be made considering the shape of the molded product, the life of the mold, the type of plastic, etc., after weighing the pros and cons. We are frequently asked which structure is good, and our answer is that which type is good or bad is determined based on the preconditions for selection.

장단점을 비교 한 후에 성형품의 모양, 금형의 수명, 플라스틱의 종류 등을 고려하여 통합구조 또는 분할구조를 사용할지 여부를 결정해야한다. 우리는 어떤 구조가 좋은지 자주 묻고, 우리의 답은 어떤 유형이 좋든 나쁘든 선택을 위한 전제조건에 따라 결정되는 것이다.

G 038 Method of Connecting the Cavity and Core

G-038 캐비티와 코어를 연결하는 방법

Usually the cavity and the core are connected to the mold plate of the mold base, but some typical of the other fastening methods are shown below.

Explanatory diagrams of the examples of four types are shown in Fig. 1.

일반적으로 캐비티와 코어는 몰드베이스의 형판에 연결되지만 몇 가지 고전적인 고정 방법은 아래에 나와 있다.

4가지 유형의 설명에 대한 그림은 1과 같다.

[Fig. 1] Methods of connecting the cavity and the core

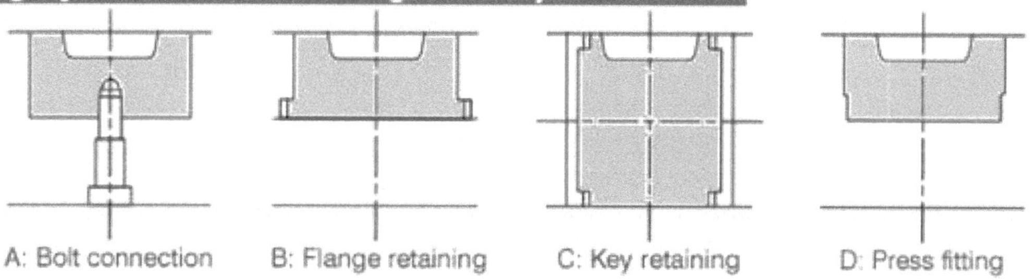

1. Bolt connecting

Bolt connecting is the method used frequently in large molds in which there is enough space to provide screw holes and bolt fixing holes in the cavity and the core. It is possible to fasten strongly because of the screw. It is possible to consider the method of tightening the bolt from the parting surface side or the method of bolt tightening from the side of the mold plate back surface. The direction of bolt tightening is selected suitably depending on the conditions. For the screw, usually metric coarse thread is used.

1. 볼트 연결

볼트 연결은 캐비티와 코어에 나사 구멍과 볼트 고정 구멍을 제공할 수 있는 충분한 공간이 있는 대형 금형에서 자주 사용되는 방법이다. 나사로 인하여 강력한 체결이 가능하다. 파팅면 쪽에서 볼트를 조이는 방법 또는 형판 뒤쪽 면으로부터 볼트를 체결할 수 있다. 볼트 조임 방향은 조건에 따라 적절히 선택할 수 있다. 일반적으로 미터나사가 사용된다.

2. Flange retaining

Flange retaining is the method used when sufficient space is not available for provided screw holes and bolt fixing holes in the cavity and core. Usually, the two methods of double side flange and single side flange are used.

The amount of catching by the flange is about 1 to 2 ㎜, and the height of the flange is about 3 to 10 ㎜, and the selection is made depending on the size of the core.

2. 플랜지고정

플랜지 고정법은 캐비티와 코어에 나사 구멍과 볼트 고정 구멍을 위한 공간이 없는 경우 사용되는 방법이다. 일반적으로 두 측면 플랜지와 단일 측면 플랜지의 두 가지 방법이 사용된다.

플랜지가 걸리는 양은 약 1 내지 2 ㎜이고, 플랜지의 높이는 약 3 내지 10 ㎜이며, 코어의 크기에 따라 선택된다.

3. Key retaining

Key retaining is the method used when fastening several components in which case it is difficult to use the flange retaining or the bolt fastening methods. The shape of the key can be that of the bracket character "]", straight key, round key, etc. It is necessary to consider the dimensions of key mating.

3. 키 고정

키 고정법은 여러 구성 요소를 고정 할 때 사용되는 방법으로 플랜지 고정 또는 볼트 체결 방법을 사용하기가 어려울 경우에 사용된다. 키 모양은 브래킷 문자 "]", 스트레이트 키, 라운드 키 등이 있다. 키 결합의 치수를 고려할 필요가 있다.

4. Press fitting

Press fitting is a mold fastening method that is used very frequently in the case of large molds. The external dimensions of the cavity and core are machined so that they fit tightly in the pocket carved from the parting surface side, and the press fitting is made by hitting with a plastic hammer, etc.
If this method is used, since the dismantling of the cavity and core becomes difficult at the time of maintenance, sometimes a hitting hole is provided in the bottom surface of the pocket for dismantling.

4. 프레스 피팅

프레스 피팅법은 대형 금형의 경우에 자주 사용되는 금형 고정 방법이다. 캐비티와 코어의 외형 치수는 가공면 측에서 조각 된 포켓에 단단히 맞도록 가공되고, 프레스 피팅은 플라스틱 해머 등으로 두드려서 만들어진다.

이 방법을 사용하면 유지관리 시에 캐비티 및 코어의 해체가 곤란해지기 때문에 해체 용 포켓의 바닥면에 타격 구멍을 설치하는 경우가 있다.

G 039 Two-Stage Ejecting Structure

The two-stage ejecting structure is a structure in which it is possible to carry out the ejecting operation twice at the time of ejecting the molded article from the core.

Although ejection is done only once in the usual injection mold, since in the case of the two-stage ejecting structure it is possible inside the mold to carry out ejection two times with a time gap between them, this is a mechanism that is very convenient in cases in which the molded item cannot be taken out the mold with only one ejecting operation.

G-039 2단 취출 구조

2단 취출 구조는 코어로부터 성형품을 취출 할 때 취출 동작을 2회 수행 할 수 있는 구조이다.

통상의 사출성형에는 1회의 취출만 이루어지지만, 성형품을 1회의 취출 조작으로 금형 밖으로 꺼낼 수 없는 경우에 사용되는 2단 취출 구조는 금형 내부에서 2회의 취출을 시간차를 두고 행할 수 있기 때문에 매우 편리한 메카니즘이다.

[Fig.1] Example in which ejecting once is not convenient

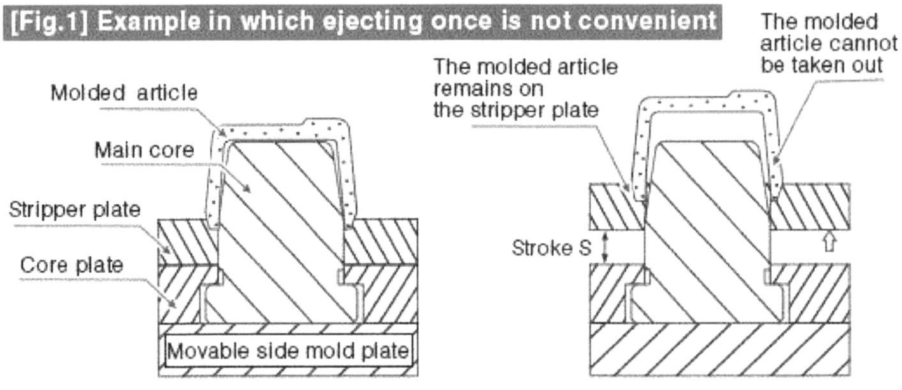

For example, in the case of the stripper plate ejecting structure such as that shown in Fig. 1, since the shape of the molded item is engraved on the stripper plate, it becomes difficult for the molded item to fall freely.

In view of this, if a two-stage ejecting structure as shown in Fig. 2 is used, it is easily possible to make the molded article fall freely.

예를 들어 그림 1에 나타낸 바와 같은 스트리퍼 플레이트 취출 구조의 경우, 성형품의 형상이 스트리퍼 상에 새겨져 있으므로, 성형품이 자유낙하가 어렵게 된다.

이러한 관점에서 그림 2에 나타낸 된 바와 같은 2단 취출 구조가 형성된다면, 성형품을 자유 낙하시키는 것이 쉬워진다.

[Fig.2] Example of the angular pin type two-stage ejecting structure

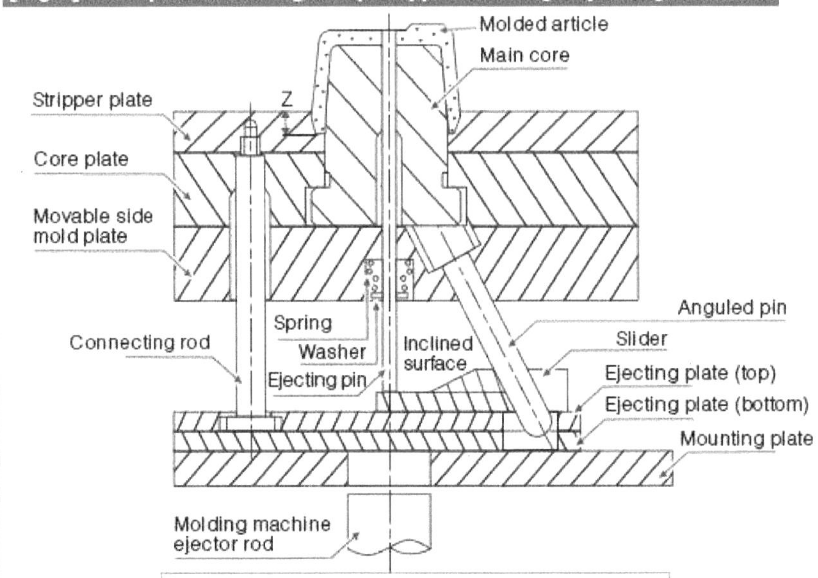

Key point: Design so that L > (Z + 2 or more)

On the other hand, in the structure shown in Fig. 3, a technique has been used in which a runner ejecting plate is provided within the ejector plate thereby making only the runner to be ejected earlier than the molded item. This has the effects of improving the gate cutting ability of the tunnel gate and of reducing the cutting scrap.

한편, 그림 3에 나타낸 구조에서는 이젝터 플레이트 내에 런너 이젝트 판을 설치하여 금형보다 빨리 런너만을 취출하는 방법이 사용되고 있다. 이것은 터널 게이트의 게이트 절단 능력을 향상시키고, 절단 스크랩을 감소시키는 효과를 갖는다.

[Fig.3] Example of a runner ejecting plate type two-stage ejecting structure

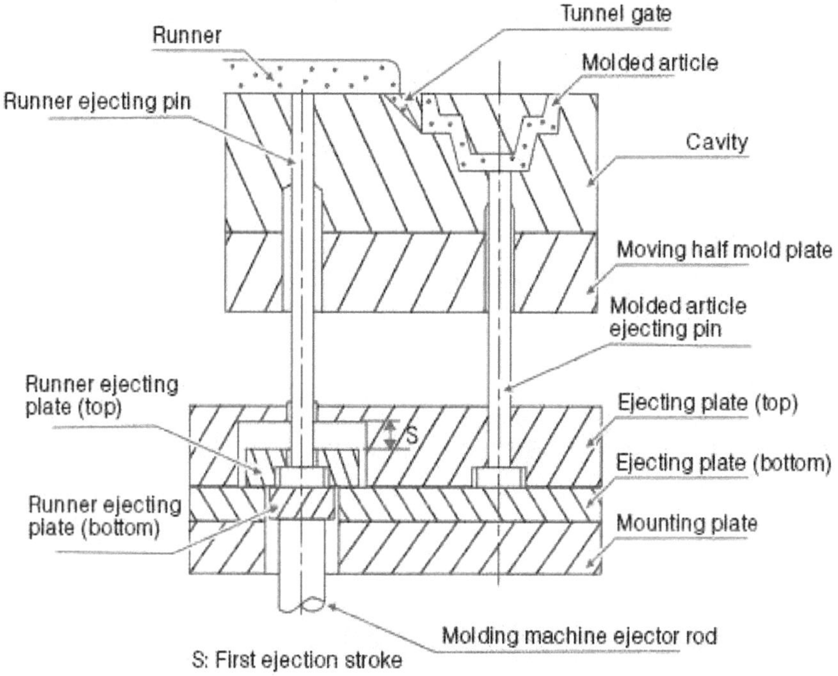

Apart from this, it is also possible to carry out two-stage ejection using other mechanisms. This will be a useful hint for solving problems when techniques have to be devised for the molded article ejection, runner, cutting of the gate, etc.

이 외에도, 다른 메커니즘을 사용하여 2단 배출을 수행하는 것도 가능하다. 성형품 취출, 런너, 게이트 절단 등의 기술을 개발해야 하는 경우의 문제를 해결하는 데 유용한 힌트가 된다.

G 040 Introduction to Products~Spiral Baffle Plates

In the molds used for plastic injection molding from now on, molding in a stable manner maintaining the shape of the molded item accurately is considered as the minimum specification, and in addition, the evaluation of the mold will vary depending on how short the molding cycle can be made.

Normally, the cooling process is the most important among the processes of the molding cycle. In order to shorten the time of the cooling process, it is very important to remove the heat efficiently from the cavity surface after the filling of the molten plastic is completed and to quickly reduce the surface temperature of the molded item.

Although flow paths are provided in a mold for passing cooling water (or cooling oil), it is not necessarily possible that effective cooling is obtained using only simple cooling holes. In order to make the coolant (liquid) act effectively, a very effective means is to take some measures to make the time and area over which the coolant comes into contact with the heat generating part larger. A very common measure is a baffle plate. The leading part of the coolant that hits against the baffle plate flows inside the cooling hole along the baffle plate and removes the heat.

A spiral baffle plate increases the probability of the coolant contacting the inside surface of the cooling holes because the coolant (liquid) flows inside the cooling hole while rotating through a spiral flow path, and hence it is possible to cool the mold more effectively.

G-040 제품 소개 ~ 스파이럴 배플 플레이트

사출성형에 사용되는 금형은 성형품의 형상을 정확하게 유지하면서 안정된 성형이 최소 사양으로 고려되며, 게다가 금형의 평가는 성형 사이클을 얼마나 단축 할 수 있는지에 따라 달라진다.

일반적으로 냉각 공정은 성형 사이클 공정 중에서 가장 중요하다. 냉각 공정의 시간을 단축하기 위해서 용융수지의 충전이 완료된 후에 캐비티 표면으로부터 열을 효율적으로 제거하고, 성형품의 표면 온도를 신속하게 낮추는 것이 매우 중요하다.

금형 내부는 냉각수(또는 냉각유)를 통과시키기 위한 냉각 채널이 설치되어 있지만, 단순한 냉각 구멍만으로는 효과적인 냉각을 얻을 수 없다. 냉매(액체)를 효과적으로 작용시키기 위해서 냉매가 발열 부품과 접촉하는 시간 및 면적을 크게 하는 것이 효과적인 방법이다. 가장 보편적인 방법은 배플 플레이트를 적용하는 것이다. 배플 판에 부딪혀 흐르는 냉매의 가장 앞부분은 배플 플레이트를 따라 냉각 구멍을 흐르면서 열을 제거한다.

스파이럴 배플 플레이트는 냉매(액체)가 나선형 유로를 통해 회전하면서 냉각 구멍의 내부로 흐르기 때문에 냉매가 냉각 구멍의 내부 표면과 접촉 할 확률을 증가 시키므로 금형을 보다 효과적으로 냉각할 수 있다.

In addition, one of the features of this product is that it can be easily assembled and adjusted. Since it is made of nylon plastic (with 30% glass fibers), it can be easily cut to the necessary length at the assembling site to match the depth of the cooling water hole. In addition, it is possible to easily remove the baffle plates when dismantling the mold, and even removing the water stains can be easily done.

Since it is quite difficult to shorten the cooling cycle with only one definitive means, it is better to carry out improvements steadily while adding several small measures one by one. As one such measure, please try using the spiral baffle plates.

또한, 이 제품의 특징 중 하나는 쉽게 조립하고 조정할 수 있다는 것이다. 나일론 플라스틱(30% 유리 섬유 함유)으로 제작되어 있으므로 냉각수 구멍의 깊이에 맞추어 조립하고, 현장에서 필요한 길이로 쉽게 절단할 수 있다. 또한, 금형을 분해 할 때 배플 플레이트를 쉽게 제거 할 수 있으며, 물 얼룩의 제거도 쉽게 할 수 있다.

오직 하나의 확실한 방법으로만 냉각 사이클을 단축하는 것은 매우 어렵기 때문에 여러 개의 작은 방법을 하나씩 추가하면서 지속적으로 개선을 수행하는 것이 좋다. 그러한 하나의 방법으로 스파이럴 배플 플레이트를 사용하자.

Spiral Baffle Plates Catalog No. WRCA / WRCT / WRCB

G 041 Bending of Core Pins due to Injection Pressure

The basic equations for calculating the bending deformation of core pins due to the injection pressure are explained here.

In injection molding, since a high filling pressure acts inside the cavity, thin and long mold parts such as core pins can get deformed or may even cause breakage accidents. The pressure acting on a core pin is different in different cases depending on the flow pattern of the molten plastic, the gate placement, etc., and hence actual calculations of the accurate strength are quite complex. Therefore, usually, the state of action of the pressure is approximated (simplified) and only basic calculations are made. The basic method of calculating the deformation (bending) of a core pin is explained below.

The maximum amount of deflection (δmax) in a cantilever beam structure is calculated using the following equations.

1. When a concentrated load acts upon the tip of the core pin

$$\delta \max = \frac{WI^3}{3EI}$$

Here, δmax is the maximum amount of bending (cm), W is the concentrated load (kgf), E is the longitudinal elastic modulus (kgf/cm^2), and I is moment of inertia of area(cm^4).

G-041 사출압력에 의한 코어 핀의 굽힘

사출압력에 의한 코어 핀의 휨 변형을 계산하기 위한 기본 방정식을 여기에 설명한다.

사출성형시에 캐비티 내부에 높은 충전압력이 작용 때문에 코어 핀 같은 가늘고 긴 부품은 변형되거나 파손 사고가 발생할 수 있다. 코어 핀에 작용하는 압력은 용융수지의 유동 패턴, 게이트 배치 등에 따라 다른 경우가 있기 때문에 정확한 강도의 실제 계산은 매우 복잡하다. 따라서 일반적으로 압력의 작용 상태는 근사화(단순화)되며, 기본 계산만 수행된다. 코어 핀의 변형(굽힘)을 계산하는 기본적인 방법을 아래에서 설명한다.

캔틸레버 보 구조에서 최대 변형량(δmax)은 다음 식을 사용하여 계산된다.

1. 코어 핀의 선단에 집중 하중이 작용했을 때

$$\delta \max = \frac{WI^3}{3EI}$$

여기서, δmax은 최대 굽힘량(cm)이며, W는 집중 하중(kgf)이며, E는 종탄성 계수(kgf/cm^2)이고, I는 단면의 관성 모멘트(cm^4)이다.

2. When a uniformly distributed load is assumed to act on the side surface of the core pin

$$\delta max = \frac{WI^4}{8EI}$$

Where, W is the uniformly distributed load (kgf/cm^2) .

In actuality, in the periphery of the core pin, since the molten plastic flows around instantaneously, it is considered rare that the pressure acts simply in only one direction. In the case of a thin and long core pin, etc., since the pressure may act during the process of (1) or (2) depending on the gate position, it is possible to carry out the basic calculations by substituting the data in the above equations.

2. 균일 분포 하중이 코어 핀의 측면에 작용하는 것으로 가정 할 때

$$\delta max = \frac{WI^4}{8EI}$$

여기서, W는 균일 분포 하중(kgf/cm^2)이다.

실제로 코어 핀의 주변에 용융수지가 순간적으로 주위에 흐르기 때문에 압력이 단순하게 한 방향으로만 작용한다고는 생각되지는 않는다. 얇고 긴 코어 핀 등의 경우에 게이트의 위치에 따라 압력이 (1) 또는 (2)의 경우가 작용하기 때문에 위의 식에 데이터를 대입하여 기본적인 계산을 수행 할 수 있다.

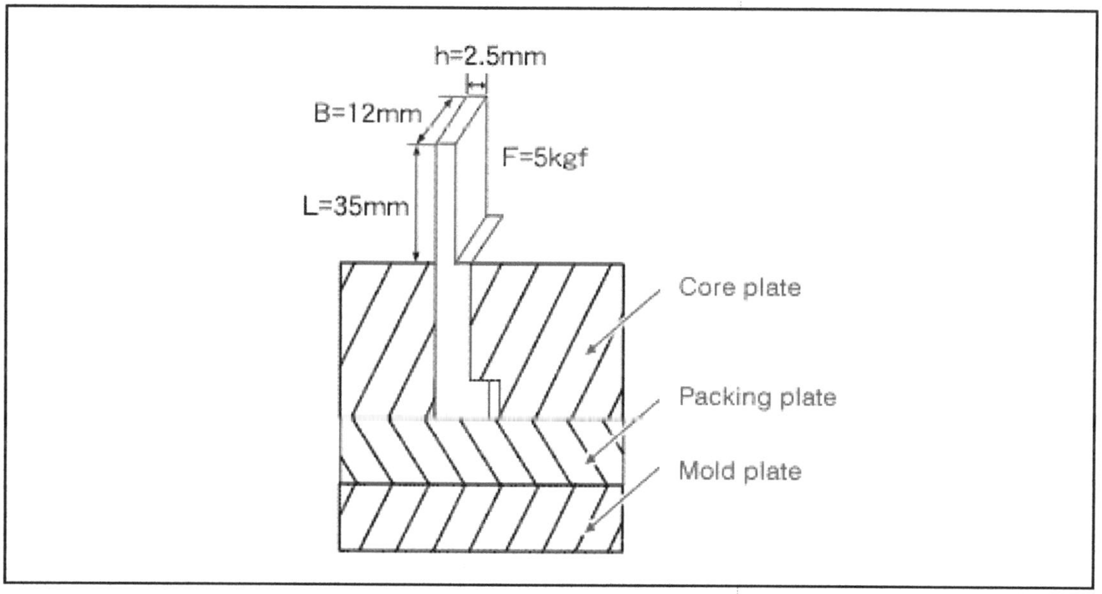

G 042 Importance of Thermal Insulation Plate

Thermal insulation plates are very important components for stabilizing the temperature of the mold or injection molding and for realizing energy savings for temperature maintenance. These are used as mandatory components in the injection molding of engineering plastics or super engineering plastics.

In general, the methods of using thermal insulation plates are the following two.

1) When using by fixing to the platen of the injection molding machine.
2) When using by fixing at the back of the mold mounting plate

The following items are the criteria for selecting thermal insulation plates.

1. Heat resistance temperature

As a guideline, the following recommended usage temperatures can be used.

- Recommended usage temperature of 100℃ or less
- Recommended usage temperature of 180℃ or less
- Recommended usage temperature of 220℃ or less
- Recommended usage temperature of 400℃ or less
- Recommended usage temperature of 500℃ or less

G-042 단열판의 중요성

단열판은 금형 또는 사출성형의 온도 안정화와 온도 유지를 위한 에너지의 절감을 위해서 매우 중요한 구성 요소이다. 이들은 엔지니어링 플라스틱 또는 슈퍼 엔지니어링 플라스틱의 사출성형에서 필수 구성 요소로 사용된다.

일반적으로 단열판을 사용하는 두 가지 방법은 다음과 같다.

1) 사출성형기의 플레이트에 고정하여 사용
2) 금형 설치판의 뒷면에 고정하여 사용

단열판을 선택하는 기준은 다음과 같다.

1. 내열 온도

지침으로 다음 권장 사용 온도를 사용할 수 있다.

- 100°C 이하의 권장 사용 온도
- 180°C 이하의 권장 사용 온도
- 220°C 이하의 권장 사용 온도
- 400°C 이하의 권장 사용 온도
- 500°C 이하의 권장 사용 온도

2. Material

The material is related to the recommended usage temperature and the strength of compression.

The following are the types of materials used.

- Cotton cloth + phenol plastic (Bakelite)
- Craft paper + phenol plastic (Bakelite)
- Glass fiber + silicate powder
- Glass fiber + ultra heat resistant epoxy resin
- Glass fiber + phosphate powder
- Glass fiber + borate powder

2. 재질

재료는 권장 사용 온도와 압축 강도와 관련이 있다.

다음은 사용 된 재료의 유형이다.

- 면직물 + 페놀 플라스틱(베이클라이트)
- 공예용 종이 + 페놀 플라스틱(베이클라이트)
- 유리 섬유 + 규산염 분말
- 유리 섬유 + 내열성 에폭시 수지
- 유리 섬유 + 인산염 분말
- 유리 섬유 + 붕산염 분말

H. 금형제작

CHAPTER H

H 001 Key Aspects of Textured Finishing

Since the textured surface finishing of molds for plastic injection molding is done after completing all of the mechanical forming operations and polishing work, if the appearance is not equal to the desired quality, it is necessary to correct it, or if the damage is so high as to make correction impossible, it is necessary to prepare the mold again. In order to complete the mold within the budget according to the schedule that was planned, it should be understood that textured finishing is a final process that has a high risk. In order to reduce the risks caused by defects in textured surface finishing, it is very important to pay attention to the following aspects.

1) The cavity surface which has to be texture finished is polished carefully using sand paper or abrasive powder thereby removing micro cracks or deformed surface layers caused by machining.
2) For the cavity material, select a steel that has a low probability of having material defects such as voids (air bubbles), inclusion of impurities, inclusion of carbides, etc.
3) Do not carry out texture finishing on parts that were repaired by welding (because it will cause striations).
4) Make the heat treatment of the steel, the cutting direction, and the rolling direction as identical as possible.
5) When processing the side surface of the cavity, set the draft as large as possible.

H-001 표면 다듬질의 중요 항목

플라스틱 사출금형의 표면 다듬질 처리는 기계적 가공 작업과 연마 작업을 모두 완료 한 후에 이루어지므로 외관이 원하는 품질과 동일하지 않은 경우, 이를 보정 할 필요가 있거나, 손상이 너무 커서 보정을 불가능하게 하는 경우에는 다시 금형을 준비 할 필요가 있다. 계획된 일정에 따라 예산 내에서 금형을 완료하려면 표면 처리가 위험성이 높은 최종 과정이라는 것을 이해해야한다. 표면 처리 문제로 인한 위험을 줄이려면 다음 항목에 주의하는 것이 매우 중요하다.

1) 다듬질이 이루어지는 캐비티 면은 샌드 페이퍼 또는 연마제를 사용하여 조심스럽게 연마하여 기계 가공으로 인한 미세 균열 또는 변형 된 표면층을 제거한다.
2) 캐비티 재료는 기포, 불순물 포함, 탄화물 등의 결함이 적은 강을 선정한다.
3) 용접으로 수리한 부분은 texture finishing을 하면 안된다. (줄무늬가 생기기 때문에).
4) 강재의 열처리, 절단 방향, 압연 방향을 가능한 동일하게 한다.
5) 캐비티의 측면을 가공 할 때, 가능한 한 크게 각도를 설정한다.

6) When carrying out processing on the side surface of the cavity, make the wall thickness of the molded product thick and intentionally make the shrinkage large.
7) Adopt a cooling structure that makes it easy to carry out temperature control of the cavity surface, and use a cartridge heater structure.
8) In some cases, a better finish is obtained if the processes are changed so that machining operations are made after texture finishing.
9) Since the visual appearance of the textured finish even changes depending on the type of the molded material, coloring, and amount of glass fibers mixed in, etc., select the type and depth of a textured finish taking into account the importance of past data.

6) 캐비티 측면에서 가공 할 때는 성형품의 살 두께를 두껍게 하고, 의도적으로 수축을 크게 한다.
7) 캐비티 표면의 온도 제어를 용이하게하는 냉각 구조를 채택하고, 카트리지 히터 구조를 사용한다.
8) 경우에 따라서는 가공이 바뀌면 가공이 완료된 후에 더 좋은 다듬질 면을 얻을 수 있다.
9) 성형된 재료의 종류, 착색 및 혼합 된 유리 섬유의 양 등에 따라 다듬질의 시각적 외관이 달라지기 때문에 과거 데이터의 중요성을 고려하여 다듬질의 유형과 깊이를 선택한다.

H 002 Improvements in the Ease of Maintenance of Molds

As mass production is continued using molds for plastic injection molding, the gas components generated from the plastic, burn or moisture in the atmosphere gets accumulated on the surfaces of the metal parts or in the gaps between the nested division parts thereby becoming the causes of molding defects. Therefore, it is necessary to dismantle and clean the mold periodically. The disassembling and cleaning of molds is generally done according to the following procedure.

H-002 금형 유지보수의 편의성 개선

금형을 이용하여 대량 생산을 계속함에 따라서 플라스틱으로부터 발생하는 가스 성분, 타거나 또는 대기 중의 수분이 금속 부품의 표면에 축척되거나, 중첩 된 분할 부품 사이에서 성형불량의 원인이 될 수 있다. 따라서 주기적으로 금형을 분해하고 청소해야한다. 금형의 분해 및 청소는 일반적으로 다음 절차에 따라 수행된다.

Disassembling the mold ↓ Cleaning and removing rust from the mold parts ↓ Assembling the mold ↓ Verifying the operation of the mold

금형 분해 ↓ 청소 및 금형 부품의 녹 제거 ↓ 금형 조립 ↓ 금형의 작동 확인

In order to make maintenance easy, it is wise to incorporate techniques at the time of designing the mold. Some ideas for this are given below.

유지 보수를 쉽게 하기 위해 금형설계 시 기술을 통합하는 것이 현명하다. 이에 대한 몇 가지 아이디어가 아래에 나와 있다.

1) Improving the accuracy of the reference surface.
2) Providing draft in the surroundings of the cavity block.
3) Providing draft in the bottom part of the excavated pocket
4) Chamfering the periphery of the bottom surface of the cavity
5) Setting auxiliary holes and auxiliary grooves for disassembling and assembling
6) Setting positioning parts (key, knock pin, etc.)
7) Adding a lubrication structure
8) Techniques for a part number assignment rule
9) Preparation of a mold maintenance manual
10) Techniques for a gas venting structure

1) 기준면의 정확도 향상
2) 캐비티 블록의 주위에 구배를 부여
3) 포켓 가공된 바닥 부분에 구배를 부여
4) 캐비티 바닥면의 주변부 모따기
5) 분해 및 조립을 위한 보조구멍 및 보조홈 설정
6) 위치 결정 부품(키, 노크 핀 등)
7) 윤활 구조 추가
8) 부품 번호 할당 규칙 기술
9) 금형의 유지보수 매뉴얼 준비
10) 가스배출 구조 기술

H 003 Hand Finishing Work

While the parts of plastic molds are mostly prepared by machining carbon steels, after the machining work is completed, final adjustment by hand finishing will be necessary to a small or a large extent. In recent machining operations, the NC data are supplied easily by CAD or CAM, and it has become possible to carry out relatively easily the machining operations at high accuracies due to the development of CNC machines or tools. However, it is necessary to carry out the final finishing by experienced and skilled hands, and since their automation is difficult, mastering this requires considerable expertise.

The details of hand finishing work are given below.

Details of hand finishing work:

Filing	Lapping	Assembling	Scraping
Marking	Drilling	Reaming	Thread cutting
Slicing	Shaving		

Measuring instruments used:

Calipers	Micrometer	Dial gauge	Block gauge
Pin gauge	Metal scale	Long metal scale	
Calipers	Metallurgical microscope		

H-003 손 다듬질 작업

사출금형 부품은 주로 탄소강을 가공하여 준비되지만, 가공 작업이 완료된 후에는 손 다듬질을 통하여 부품을 약간 작거나 크게 최종 조절할 필요가 있다. 최근의 가공 작업에는 NC 데이터가 CAD 또는 CAM을 통해 쉽게 공급되며, CNC 기계 또는 공구의 개발로 인해 높은 정확도로 가공 작업을 비교적 쉽게 수행 할 수 있게 되었다. 그러나 숙련 된 기술자가 최종 다듬질 작업을 수행해야하며, 자동화가 어려우므로 이를 숙련하기 위해서 상당한 전문 지식이 필요하다.

손 다듬질 작업의 세부 사항은 다음과 같다.

손 다듬질의 세부 작업 :

다듬질, 랩핑, 조립, 스크랩핑, 마킹, 드릴링, 리밍, 나사선 깎기, 슬라이싱, 셰이빙

측정 장비의 사용:

캘리퍼스, 마이크로미터, 다이얼 게이지, 블록 게이지, 핀 게이지, 금속 눈끔자, 긴 금속 눈끔자, 금속 현미경

In order to learn hand finishing, the most definite method is to learn by working with an experienced technician having an accurate knowledge (a person having a special technician certificate or a Grade 1 technician certificate). Although finishing can be done to some extent by looking at and copying an experienced technician, if it is not possible to work with the correct knowledge and using the correct procedure, it is difficult to carry out precision work or work that is highly paid.

손 다듬질을 배우는 가장 정확한 방법은 정확한 지식을 가진 숙련 된 기술자(특수 기술자 또는 1급 기술자 인증서를 가진 사람)와 협력하여 학습하는 것이다. 숙련 된 기술자를 보고 따라하여 어느 정도 마무리 작업을 수행 할 수 있지만, 정확한 지식과 올바른 절차로 작업이 불가능하다면 정밀 작업이나 높은 보수를 받기는 어렵다.

H 004 Filing Work

H-004 줄 작업

Correct work procedure is very important for carrying out filing work appropriately. The key points of filing work are the following.

올바른 작업 절차는 적절하게 줄 작업을 수행하는 데 매우 중요하다. 줄 작업의 요점은 다음과 같다.

1. Posture during filing

When filing, first the posture should be correct. Otherwise, it will not be possible to do accurate and precision work.

1) Adjust the height of the work to be at the height of your elbows and fix it firmly using a vice, etc.
2) Stand in front of the work and place your body so that the tip of the file is at the center of the work.
3) Place your right foot along the center line of the work, and tilt the right foot so that is at an angle of 70° to 80° to the center line.

1. 줄 작업 시의 자세

줄 작업 시에 먼저 자세가 정확해야 한다. 그렇지 않으면 정확하고 정밀한 작업을 수행 할 수 없다.

1) 작업 물의 높이를 팔꿈치 높이로 조정하고 바이스 등을 사용하여 단단히 고정.
2) 작업 물 앞에 서서 줄의 끝 부분이 작업 물의 중앙에 오도록 몸을 위치시킴.
3) 오른발을 작업 물의 중심선을 따라 놓고, 오른발을 중심선에 대해 70°~80°가 되도록 기울일 것.

4) Open your left foot and step it forward in the direction of the work by half a step, and place the step so that the tip of the foot is about 200 ㎜ to 300 ㎜ distant from the work.
5) Relax the arm muscles, and make fine adjustment of the body and legs so as to be in a position where the file can be moved forwards and backwards lightly with respect to the work.

4) 왼발을 열고, 작업물 쪽으로 반 발짝 전진하고, 발끝이 작업 물에서 약 200 ~ 300 ㎜ 떨어져 있도록 스텝을 위치시킴.
5) 팔의 근육을 이완시키고, 작업물에 대해 줄을 앞뒤로 가볍게 움직일 수 있는 위치에 있도록 몸과 다리를 미세 조정.

2. Filing work

There are three types of filing work.

1) Straight movement method : This is the most comm on method of moving the file in the straight forward direction. The finished surface becomes clean and neat.
2) Oblique movement method : This is the method of moving the file in a direction that is moving form the bottom left to the right. This method is suitable for rough scraping because the amount cut is large.
3) Combined movement method : This method is suitable for long objects with small widths.

2. 줄 작업

줄 작업은 세 가지 유형

1) 직선 이동 방법 : 줄을 직진 방향으로 이동시키는 가장 일반적인 방법이다. 완성 된 표면이 깨끗하다.
2) 경사 이동 방법 : 줄을 왼쪽 아래에서 우측으로 이동시키는 방법이다. 이 방법은 절삭양이 많기 때문에 거친 작업에 적합하다.
3) 조합 이동 방법 : 이 방법은 폭이 작은 긴 물체에 적합하다.

3. Removing the filing dust from the file

Since the cut dust gets clogged in the grooves of the file, frequently the grooves of the file must be cleaned during filing. Brush away the dust clogged in the grooves of the file using a wire brush or the dust can also be blown off using a compressed air blast. Another technique of preventing clogging of the file grooved by filing dust is to coat the file before working using charcoal or chalk.

3. 줄의 절삭 잔유물을 제거

절삭된 잔유물로 인하여 줄의 홈이 막히기 때문에 줄의 홈을 자주 청소해야한다. 와이어 브러쉬를 사용하여 줄 홈을 막은 절삭 잔유물을 털어 내거나 압축 공기를 사용하여 날려 버릴 수도 있다. 절삭 잔유물로 줄 홈이 막히는 것을 방지하는 또 다른 기술은 숯이나 분필로 작업하기 전에 줄을 코팅하는 것이다.

H 005 Scraping Work

Scraping work is that of scraping off very fine depressions on the polished surface using a tool called a "scraper". Such scraping operation is done on the sliding surface of the bed of a machine or on the surfaces of moving machine parts.

The purposes of scraping operations are the following.

1) To make the lubricating oil stay evenly on the surface.
2) To reduce the friction resistance by making the contact area small.
3) To reduce the generation of friction heat by making the contact area small thereby reducing the friction resistance.
4) To suppress thermal expansion by reducing the generation of friction heat.
5) To suppress the increase in friction resistance due to thermal expansion.

There are the following types of scraping work,

1) Flat scraping
2) Hook scraping
3) Cant scraping

H-005 스크래핑 작업

스크래핑 작업은 "스크레이퍼"라고 불리는 도구를 사용하여 광택면의 매우 미세한 함몰부를 깎아내는 작업이다. 이러한 스크래핑 작업은 스크래핑 기계의 슬라이딩 배드부 또는 이동하는 기계 부품의 표면에서 수행된다.

스크래핑 작업의 목적은 다음과 같다.

1) 윤활유를 표면에 균일하게 도포
2) 접촉 면적을 작게 함으로써 마찰 저항 감소
3) 접촉 면적을 작게하여 마찰열을 감소시켜 마찰 저항을 감소
4) 마찰열의 발생을 억제하여 열팽창을 억제
5) 열팽창에 의한 마찰 저항의 증가를 억제

스크래핑 작업의 종류

1) 플랫 스크래핑
2) 훅 스크래핑
3) 칸트 스크래핑

When carrying out scraping work, it is very important to maintain a clean environment in the work place. It is essential to remove all the time any dust, cutting shreds, sand, etc., using a vacuum cleaner, etc. In addition, when carrying out scraping work, an appropriate cutting oil should be selected, and the work should be carried out while applying it in suitable quantities.

Red lead is used for carrying out scraping work. This is a technique for carrying out scraping work uniformly while checking the scraped surface.

As explained above, scraping work is a precision finishing operation that is to be made mandatorily on the sliding surface, etc., of a precision metal working machine. The reason why Japanese metal working machines are of high precision and with very little deviation is the presence of scraping work.

Although there are very limited examples of carrying out scraping work in forming the parts of molds for plastic molding, depending on the usage this some times has very useful functions. This is the ideal method for preventing the biting of slide cores or moving parts.

The history of the design and preparation of molds for plastic molding is still quite shallow being only about 50 years. It is possible to obtain useful effects by observing well and applying the know how used in other precision machine technologies.

스크래핑 작업을 수행 할 때 작업장을 깨끗하게 유지하는 것이 매우 중요하다. 진공 청소기를 사용하여 먼지, 절삭 파편, 모래 등을 항상 제거해야한다. 또한 스크래핑 작업을 수행 할 때는 적절한 절삭유를 선택해야하며 적절한 양으로 작업해야한다.

홍연광(紅鉛鑛)은 스크래핑 작업을 수행하는 데 사용된다. 이것은 스크래핑 된 표면을 확인하면서 스크래핑 작업을 균일하게 수행하는 기술이다.

앞서 기술한 바와 같이, 스크래핑 작업은 정밀 금속 가공 기계의 슬라이딩 표면 등에 강제적으로 행해지는 정밀 가공 작업이다. 일본의 금속 가공 기계가 정밀도가 높고 편차가 거의 없는 이유는 스크래핑 작업이 존재하기 때문이다.

사출금형 부품을 제작할 때 스크래핑 작업을 수행하는 사례는 매우 제한되어 있지만 사용법에 따라 매우 유용한 기능을 가지고 있다. 이것은 슬라이드 코어 또는 움직이는 부품의 뜯김을 방지하는 이상적인 방법이다.

플라스틱 성형을 위한 금형설계 및 준비의 역사는 약 50 년으로 오래 되지 않았다. 잘 관찰하고, 다른 정밀 기계 기술에 사용 된 노하우를 적용하여 유용한 효과를 얻을 수 있다.

H 006 Mold Assembling Work

The work of assembling a plastic injection mold is a process step that is always necessary whenever a new mold is prepared. It is very common to carry out the assembly of a mold using the following work procedure.

1) Confirming the component count of the mold parts
2) Confirming the procedure of assembling the mold
3) Provisional assembly of the mold parts
4) Adjusting the operations
5) Final assembly of the mold
6) Confirming the operation in a molding machine
7) Adjusting the defective locations (if necessary, dismantle and reassemble the mold)

In the process step of assembling a mold, since the results of all operations such as design, machining, finishing, surface treatment, etc., all are collected together, this is also a process step in which any human mistakes or insufficiencies in the respective processes such as design mistakes, machining mistakes, part ordering mistakes, etc., all come out. It is necessary to detect speedily such problematic locations and to arrange for design modifications or machining of the mold. Human mental work or machine processing time are physically required for design modifications or part machining of molds. Whether appropriate and accurate judgment and speedy follow up actions can be done smoothly or not determines the trial schedule of the mold and the amount of time taken to a large extent.

H-006 금형 조립 작업

사출금형을 조립하는 작업은 새로운 금형을 준비 할 때마다 항상 필요한 공정이다. 다음 작업 절차를 이용하여 금형을 조립하는 것이 일반적이다.

1) 금형 부품의 개수 확인
2) 금형 조립 순서 확인
3) 금형 부품의 임시 조립
4) 작동 조정
5) 금형의 최종 조립
6) 성형기에서의 동작 확인
7) 결함 위치 조정 (필요한 경우 금형 분해 및 재조립)

금형을 조립하는 단계에서 설계, 가공, 다듬질, 표면 처리 등의 작업이 모두 종합되는 작업으로, 설계 오류, 가공 오류, 부품 주문 오류, 전체 작업 등의 각 공정에서 설계자 및 가공 엔지니어의 오류나 문제점이 나타나는 단계이기도 하다. 이러한 문제가 있는 위치를 신속하게 감지하고, 금형의 설계 수정 또는 가공을 준비해야한다. 엔지니어의 정신적 작업이나 기계 가공시간은 금형의 설계 변경이나 부품 가공에 필요하다. 적절하고 정확한 판단과 신속한 후속 조치가 원활하게 수행 될 수 있는지 여부 또는 금형의 시험 일정 및 연장되는 시간을 정하지는 않는다.

Here, since the parts or data pass through a large number persons and processes until the mold is completed, there will be some inadvertent mistakes in information transmission or misunderstandings by individuals. Although it is important to take actions so that these do not occur, and there are some persons who manage saying "this should never happen according to psychological theory", but that is not an advisable approach. It is very important to have an organization and a command chain system for quickly meeting any problems that have occurred, and this is a realistic approach for such problems.

Let us give you some individual advices for the assembly process.

여기서, 부품 또는 데이터는 많은 수의 사람을 거쳐 금형이 완료 될 때까지 처리되기 때문에 개인에 의한 정보 전송이나 오해에 의한 우연한 실수가 있을 수 있다. 이런 일이 일어나지 않도록 하는 것이 중요하지만 "심리적 이론에 따라 이런 일은 있어서는 안 된다."라고 말하는 사람이 있다. 하지만 이는 바람직한 방법이 아니다. 발생한 문제를 신속하게 해결할 수 있는 조직 및 명령 체계를 갖추는 것이 매우 중요하며, 그리고 이것이 문제에 대한 현실적인 접근 방법이다.

어셈블리 공정에 대한 개별 조언을 제공한다.

Verifying the mold assembly procedure

- It is very important to start the assembly work only after thoroughly understanding the assembly procedure. When some work can be carried out simultaneously in parallel, the time taken can be shortened if the work is divided among two or more persons.

금형 조립 절차 확인

- 조립 절차를 완전히 이해 한 후에 조립 작업을 시작하는 것이 중요하다. 일부 작업을 동시에 병행할 경우, 작업이 두 명 이상의 사람들로 나누어져 있는 경우에는 소요되는 시간이 단축될 수 있다.

Provisional assembly:

- When tightening parts using a hexagonal socket head bolt, etc., all the screws should not be tightened in the beginning, but about four bolts should be selected that take up the maximum force, and these bolts should be provisionally tightened in a balanced manner and in the sequence of opposite bolts first, and these bolts should be tightened fully in the end when the overall position of the part has been finely adjusted. After that, by tightening the remaining bolts, it is possible to stabilize the tightening of the entire mold.

임시 조립 :

- 6각 머리 볼트 등으로 체결 할 때는 처음부터 모든 나사를 조이지 말고 최대 힘을 취할 수 있는 약 4개의 볼트를 선택하여 균형 있게 조인다. 먼저 균형 있게 조인 후에 반대쪽 볼트 순서로 조여야하며, 이 볼트는 부품의 전체 위치가 미세하게 조정된 후에 끝까지 완전히 조여져야한다. 그 후에 나머지 볼트를 조여서 금형 전체의 조임을 안정시킨다.

- It is possible to improve the efficiency of the assembly work by the use of small appliances such as steel blocks or pins, sponge, woolen cloth, rubber plates, etc.

- 스틸 블록이나 핀, 스폰지, 모직 천, 고무판 등과 같은 소형기기를 사용하여 조립 작업의 효율성을 높일 수 있다.

Adjusting the operation of the mold:

- In order to confirm the operations of the slide core, ejector pin, return pin, etc., fine adjustments of different parts should be made after thoroughly checking the entire mold. If the work is done by concentrating only on one location right from the start, the assembly will have been done "looking at a tree but not looking at the forest", and hence care should be taken because this can make the mold not to function properly.

금형 작동 조정 :

- 슬라이드 코어, 이젝터 핀, 리턴 핀 등의 동작을 확인하기 위해서 금형 전체를 철저히 점검 한 후 각 부분을 미세 조정을 해야 한다. 작업이 처음부터 하나의 위치에만 집중하여 이루어지면 조립은 "나무를 보지만 숲을 보지 못한다."와 같은 경우가 되므로 금형이 기능을 하지 못하게 될 수 있으므로 주의해야 한다.

H 007 Mold Welding Work

When a part of a component of a mold for plastic injection molding has become worn out or when a corner part has been chipped off due to hitting it against some hard object, etc., welding can be used as a method for repairing such parts. Although welding is not used in the case of precision molds because welding can result in big damage to the mold component due to high temperatures or formation of oxide layers during welding, welding is an important repairing technology in the case of large molds, or when the external appearance quality of molded products is not very severe, or for carrying out emergency repairs. In view of this, the outlines of welding are described below.

H-007 금형 용접 작업

사출금형의 부품 일부가 마모되거나, 경질 물체 등에 부딪혀서 모서리가 절단했을 경우, 용접으로 수리 할 수 있다. 용접은 고온 또는 용접 중 산화물 층의 형성으로 인해 금형 부품에 큰 손상을 줄 수 있기 때문에 정밀 금형의 경우에는 사용되지 않지만, 대형 금형, 성형품의 외관 품질이 그다지 중요하지 않은 경우, 응급 수리가 필요한 경우에는 중요한 수리 기술이다. 이 점을 고려하여 용접의 개요를 설명한다.

1. Types of welding

The following are the three major types of welding used for repairing molds.

1) Brazing
2) Gas welding
3) Electrical arc welding

1. 용접의 종류

다음은 금형 수리에 사용되는 용접의 세 가지 주요 유형이다.

1) 브레이징
2) 가스 용접
3) 전기 아크 용접

2. Features of the different welding methods

(1) Brazing

Brazing is that of affixing to the part to be repaired a material such as solder or silver by heating it in a flame and melting it. Since the repair will be made using a metallic material that is different from the base material which is carbon steel, the strength of the joint is not very good. However, since this type of welding work is simple, it is suitable for making emergency repairs.

(2) Gas welding

In general, this refers to oxy-acetylene welding. Oxygen and acetylene gases are supplied from two gas cylinders, the two are ignited and their flow rates are adjusted to obtain a very hot pointed jet of flame and the welding is carried out by feeding a welding rod between the part to be repaired and the flame. Although the strength of the welded joint is lower than that of electric arc welding, it can meet sufficiently the purpose of repairs. Compared to electric arc welding, since the thermal effect on the base material is small, this method is suitable for molds that are sensitive to deformations.

2. 여러가지 용접 방법의 특징

(1) 브레이징

브레이징은 납이나 은과 같은 물질을 가열하고 녹여 수리 할 부분에 부착하는 것이다. 수리는 탄소강인 기본 재료와 다른 금속 재료를 사용하여 이루어지므로 접합 강도는 그리 좋지 않다. 그러나 이 용접 작업은 간단하기 때문에 응급 수리 작업에 적합하다.

(2) 가스 용접

일반적으로 옥시 아세틸렌 용접을 말한다. 산소와 아세틸렌 가스는 2개의 가스 실린더에서 공급되고, 두 가스를 점화시킨다. 가스의 유량은 매우 뜨겁고 뾰족한 불꽃을 얻기 위해 조정되고, 용접은 수리 될 부품과 화염 사이에 용접봉을 공급하면서 이루어진다. 용접부의 강도는 전기 아크 용접보다 낮지만 수리의 목적으로는 충분하다. 전기 아크 용접과 비교하여 모재에 대한 열 부작용이 적으므로 이 방법은 변형에 민감한 금형에 적합하다.

(3) Electric arc welding

This is the method of creating heat by causing an arc using an AC current or a DC current. The following three methods are used in this type of welding.

1) Coated electrode arc welding

This arc welding is done using coated welding rods. This is used very frequently for ordinary mechanical parts.

2) Carbon dioxide gas arc welding

In this method, the welding is done in a carbon dioxide gas shield. It is possible to prevent oxidization of the welding parts.

3) Inert gas arc welding

In this method, the welding is done in an inert gas shield such as argon gas, etc. This is the most suitable method for repairing molds.

Since it is possible to prevent definitely any oxidization of the welded parts and to suppress the generation of oxides, and generation of blow holes or air bubbles is difficult even if grinding is done. However, the cost of the work becomes high, and there is the limitation that this work cannot be made unless the company has the required equipment.

(3) 전기 아크 용접

이것은 AC 전류 또는 DC 전류를 사용하여 발생시킨 아크의 열을 이용하는 방법이다. 이러한 용접에는 다음 세 가지 방법이 사용된다.

1) 코팅 전극 아크 용접

이 아크 용접은 코팅 된 용접봉을 사용하여 수행된다. 일반 기계 부품에 매우 자주 사용된다.

2) 이산화탄소 가스 아크 용접

이 용접은 이산화탄소 가스 차폐로 이루어진다. 용접 부분의 산화를 방지 할 수 있다.

3) 불활성 가스 아크 용접

이 용접은 아르곤 가스와 같은 불활성 기체 차폐물(inert gas shield)에서 수행된다. 금형 수리에 가장 적합한 방법이다.

용접부의 산화를 확실하게 방지하여 산화물의 생성을 억제 할 수 있고, 연삭을 행해도 블로우 홀이나 기포의 발생이 어렵다. 그러나 작업 비용이 높아지고 회사에 필요한 장비가 없다면 이 작업을 수행 할 수 없는 한계가 있다.

H 008 On Assembling a Square Core

H-008 사각 코어의 조립

An embedded split structure is used very frequently as the cavity and core structure. This splitting has the following purposes.

1) A part that is difficult to form by machining is divided as a different part, thereby making its machining easy. The machining cost can be reduced and it is possible to increase the machining quality.
2) It is possible to provide the air vent function of discharging air or gas from the parted surface.
3) A part that can easily break is divided in advance so that it can be replaced easily.

The following patterns can be considered as the patterns for embedded splitting.

1) Round core pin shape splitting
2) Square core shape splitting
3) Odd shape splitting

Among these, a little know-how is described here related to assembling when the splitting is made in the shape of a square core.

The following figure shows an example of splitting a square core shape. In order to insert a square core, it is necessary to machine a square shaped hole in the main core. The commonly used machining methods are - wire cutting electric discharge machining, shape carving electric discharge machining, or further dividing the main core and carrying out grinding operations.

분할 삽입 구조는 캐비티 및 코어 구조로 자주 사용된다. 이 분할 삽입 구조의 목적은 다음과 같다.

1) 기계 가공으로 제작하기 어려운 부분은 별도 부분으로 나뉘어 가공을 용이하게 한다. 가공비용을 절감 할 수 있으며, 가공 품질을 향상시킬 수 있다.
2) 분할 된 표면으로 공기 또는 가스를 배출하는 등의 공기 배출 기능을 할 수 있다.
3) 쉽게 부서 질 수 있는 부분은 사전에 분리되어 쉽게 교체 할 수 있다.

다음 패턴은 분할 삽입으로 고려 할 수 있다.

1) 라운드 코어 핀 모양 분할
2) 사각형 코어 모양 분할
3) 특별 형상 분할

그 중 사각형 코어의 형태로 분할 할 때 조립과 관련하여 약간의 노하우가 여기에 설명되어 있다.

다음 그림은 사각형 코어 분할의 예를 나타낸 것이다. 사각형 코어를 삽입하려면 주 코어에 사각형 모양의 구멍을 가공 할 필요가 있다. 일반적으로 사용되는 가공 방법은 와이어 커팅, 방전 가공, 또는 주 코어를 더 분할하고, 그라인딩 작업을 수행한다.

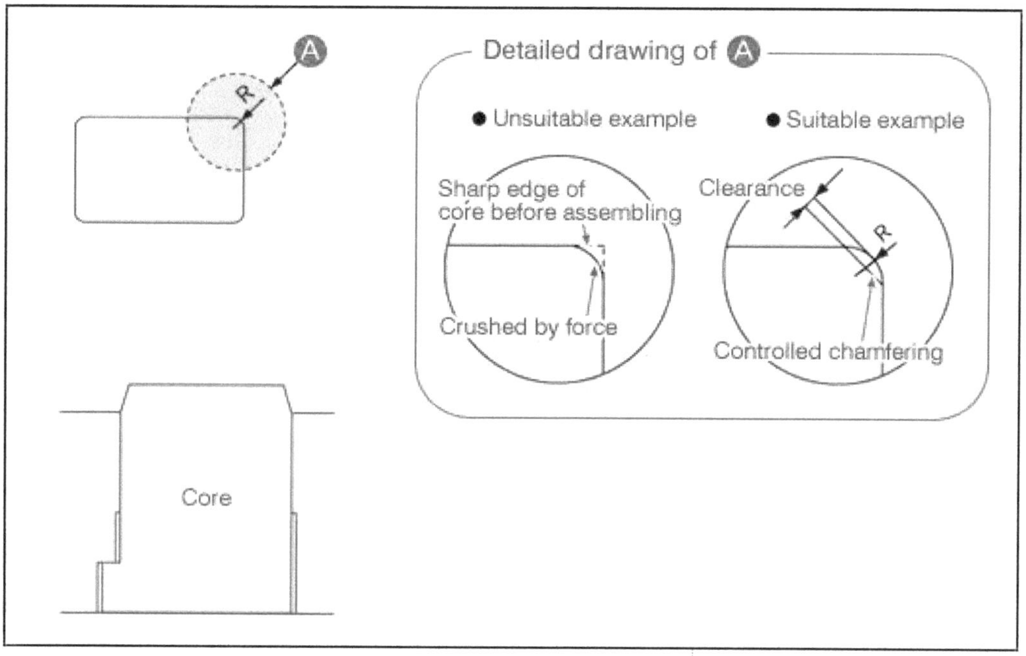

Whatever be the machining method used, a very small rounding will remain in the corner part. The size of rounding is about R0.15 ㎜ to R0.05 ㎜. If even such a fine rounding remains, if assembly is attempted with the corner of the core having a sharp edge, the assembling does not progress smoothly. If hitting with a copper ha mmer is done by force the assembly, the edge of the corner is likely to get crushed in an unstable manner.

In order to carry out smooth assembling, fine chamfering is done of the sharp edges of the core by machining or by hand finishing so as to make the edges match with the corner radius R of the square hole. If this is done, the assembling can be done smoothly. Since the magnitudes of R and C are very slightly different, although a clearance may be generated, if the clearance is about 0.03 ㎜, it can be said to be too small for any possibility of the molten plastic flowing in and creating burrs. On the contrary, it may also have the effect of discharging air or gas.

사용 된 가공 방법이 무엇이든, 모서리 부분에 아주 작은 라운드가 남는다. 라운드의 크기는 R0.15 ㎜에서 R0.05 ㎜이다. 이렇게 미세한 라운딩이 남아도 날카로운 모서리를 갖는 코어의 모서리에 조립을 시도하면 조립이 원활하게 진행되지 않는다. 구리 해머로 두들겨 조립을 강제로 수행하는 경우 모서리의 가장자리가 불안정한 방식으로 부서지기 쉽다.

매끄러운 조립을 수행하기 위해서 가공 또는 손으로 마무리하여 코너의 날카로운 모서리를 정밀하게 모따기하여 모서리가 사각 구멍의 코너 반경 R과 일치하도록 한다. 이 작업이 완료되면 원활하게 조립할 수 있다. R과 C의 크기가 아주 약간 다르므로 공차가 발생할 수는 있지만 공차가 약 0.03 ㎜이면 용융수지가 흘러 들어가 버가 발생할 수 없다. 반대로, 공기 또는 가스를 방출하는 효과는 있을 수 있다.

However, in the case of special environments such as a material like PPS having very good fluidity, it is recommended that the management of the clearance is made based on prototyping data.

그러나 유동성이 좋은 PPS와 같은 특수한 환경의 경우, 프로토 타입 데이터를 기반으로 공차를 관리하는 것이 좋다.

H 009 Mating With Nested Divisions

H-009 겹치는 부품의 끼워 맞춤 결합

"Mating" is a concept of controlling the permissible dimensional tolerances of the shape of the shaft and the shape of the hole, and the following three are the methods of mating.

"맞춤"은 축(軸) 형상 및 구멍 모양의 허용 가능한 치수 공차를 제어하는 개념이며, 다음 세 가지가 결합 방법이 있다.

(1) Gap mating

This is the method of mating in which there is always a gap in any combination of the shaft dimensional tolerance and the hole dimensional tolerance.

(1) 헐거운 끼워 맞춤

축(軸) 공차와 구멍 치수 공차의 조합에 항상 간격이 있는 결합 방법이다.

(2) Intermediate mating

This is the method of mating in which a gap may be generated or a tight mating margin may be generated depending on the combination of the shaft dimensional tolerance and the hole dimensional tolerance.

(2) 중간 끼워 맞춤

축(軸) 공차와 구멍 치수 공차의 조합에 따라 틈이 생기거나 밀착되는 맞춤 방법이다.

(3) Tight mating

This is the method of mating in which a tight mating margin is present at all times in any combination of the shaft dimensional tolerance and the hole dimensional tolerance.

(3) 억지 끼움 맞춤

축(軸) 공차와 구멍 치수 공차의 조합에 항상 억지 끼워 맞춤 여유가 있는 방법이다.

In addition, there are the two methods 'hole reference method" and "shaft reference method" of machining the hole and shaft that mate with each other of taking either the hole or the shaft as the reference and making the other match with the reference.

In the hole reference method, the dimensional tolerances of the hole are set in advance as the reference, and the dimensional tolerances of the mating shaft are made to match with the reference dimensional tolerances. In this method it is possible to realize relatively easily the desired mating by finishing the dimensions of the shaft accurately. The hole reference method is very frequently used for the mating of the cavity and core in the molds for plastic injection molding.

In the shaft reference method, the dimensional tolerances of the shaft are set in advance as the reference, and the dimensional tolerances of the mating hole are made to match with the reference dimensional tolerances. In this method, in order to machine the hole accurately to match with the shaft, a milling machine or a jig borer with a good accuracy will be required.

The mating method should be determined by the choice made by the designer of the mold. Which method is to be used for mating with which part has to be decided, and also it is necessary to decide what should be the permissible tolerances.

또한 구멍 또는 축을 기준으로 서로 맞 물리는 구멍 및 축을 가공하고 다른 하나를 기준과 일치시키는 '구멍 기준 방법'과 '축(軸) 기준 방법'의 두 가지 방법이 있다.

구멍 기준법은 구멍의 치수 공차는 미리 기준으로서 설정되고, 상대 축(軸)의 치수 공차는 기준 치수 공차와 일치하게 된다. 이 방법에는 축(軸)의 치수를 정확하게 마무리하여 원하는 끼워 맞춤을 비교적 쉽게 실현할 수 있다. 구멍 기준 방법은 플라스틱 사출성형을 위한 금형에서 캐비티와 코어의 결합에 매우 자주 사용된다.

축(軸) 기준법에는 축의 치수 공차를 기준으로 사전 설정하고 상대 구멍의 치수 공차를 기준 치수 공차와 일치시킨다. 이 방법에는 축에 맞게 구멍을 정밀하게 가공하기 위해 정밀도가 좋은 밀링 머신 또는 지그 보링기가 필요하다.

끼워 맞춤법은 금형설계자가 선택하는 것에 따라 결정된다. 어떤 방법으로 끼워 맞춤법을 결정하여야 하는지, 그리고 허용되는 공차가 어떻게 되는지 결정해야 한다.

H 010 Polishing the Cavity

After a plastic injection mold has been machined using milling, electric discharge machining, or wire cut machining, the surface of its cavity has to be finished mainly by human hands or by polishing using a machine so that the surface is made extremely smooth.

If the condition of polishing is not proper, the glossiness decreases of the transferred surface of the molded item, and the quality of the surface becomes poor. In addition, at the time that the molded item is separated from the cavity, the undulations in the surface become undercuts which may cause defects in separating the molded item from the mold.

The polishing of the cavity surface is done using a hard grinding stone or abrasive powder. Grinding stones or abrasive powders are used successively starting from those with a large grain size and ending with those a small grain size. The polishing work is done while applying cutting oil so that there is no clogging or biting.

The direction of polishing is not only one fixed direction, but the polishing direction is changed frequently such as in the left to right direction, top to bottom direction, or circular direction. Considerations are given so that the polished surface is uniform.

The following are the main polishing materials used.

H-010 캐비티의 연마

밀링, 방전 가공 또는 와이어 컷 가공을 이용하여 플라스틱 사출금형을 가공 한 후에는 캐비티 표면을 주로 사람의 손으로 마무리하거나 기계로 연마하여 표면이 매우 매끈하게 해야 한다.

연마 상태가 적당하지 않은 경우, 성형품의 전사면의 광택이 저하되어 표면의 품질이 나빠진다. 또한, 성형품이 캐비티로 부터 분리 될 때, 표면의 기복이 언더컷이 되어 금형으로부터 성형품을 분리시키는 데 있어서 결함을 유발할 수 있다.

캐비티 표면의 연마는 단단한 연마석 또는 연마제를 사용하여 수행한다. 연삭숫돌이나 연마제는 큰 입자 크기에서 작은 입자 크기로 연속적으로 사용된다. 연마 작업은 절삭유를 추가하면서 막힘이나 뜯김이 생기지 않도록 한다.

연마 방향은 한 방향으로만 가는 것이 아니라, 좌에서 우방향, 위에서 아래로, 원형 방향으로 자주 방향을 바꿔야 한다. 연마된 표면이 균일하도록 고려해야 한다.

다음은 주요 연마재이다.

Natural polishing materials	Silica stones (onyx, Arkansas stone, etc.) Corundum stones (emery, garnet) Diamond powder
Artificial polishing materials	Alumina (Al2O3) Silicon carbide (SiC) Boron carbide (BC) Artificial diamond

The granularity of polishing materials can be in the range of #10 to #3000. As the number becomes smaller, the grain becomes rougher.
(Example: The finish will be rough when polished with a grain size of #10 to #30, and the finish will be fine and smooth when polished with a grain size of #1000 to #2000.)

When using a fine grinding stone, the polishing is done by mixing olive oil or a vegetable oil to a wooden material such as maple, pine, willow, balsa, etc.

연마재의 입도는 #10에서 #3000 범위이다. 숫자가 작아지면 입자가 거칠어진다.(예 : #10에서#30의 입자 크기로 연마하면 표면이 거칠어지며, #1000에서 #2000의 입자 크기로 연마 될 때 표면이 부드럽고 매끄러워진다.

미세 연삭숫돌을 사용하는 경우, 단풍나무, 소나무, 버드나무, 발사 등의 목재에 올리브 오일이나 식물성 기름을 혼합하여 연마한다.

H 011 Hard Chromium Plating

H-011 하드 크롬 도금

The surfaces of the cavity and core are likely to be oxidized (rusted) and corroded by the volatile components from the molded material or from moisture content in the atmosphere.

Considering the mold release of the molded article, it is necessary that the surface of the cavity is carefully polished and is maintained in a shining condition.

캐비티 및 코어의 표면은 성형재료의 휘발성 성분 또는 대기의 수분 함량에 의해 산화되거나 (녹슬어) 부식되기 쉽다.

성형품의 이형성을 고려하면, 캐비티의 표면을 정성스럽게 연마하여 빛나는 상태로 유지할 필요가 있다.

Plating the surface of the cavity is used as a measure for preventing oxidization or corrosion. Hard chromium plating is a typical method of plating among the different varieties of plating.

Hard chromium plating has good wear resistance and corrosion resistance.

The hardness of the film on the surface is 700 to 1000 Hv, and the surface has a white metallic glossiness.

To carry out the plating, the work to be plated is immersed in a plating solution made of chromic acid anhydride, sulfuric acid, etc., and an electric current is passed to form an electroplated film.

The plated film is a robust passivated film of a base metal in air. Since the surface energy of the plated film is low, it has the property of being hard to be adhered to by other materials. Therefore, it can also be said that it has the property of making it easy for the plastic molded article to be released from the mold.

In order to carry out plating skillfully, it is recommended to carefully polish the part of the surface that is to be plated, thereby putting it in a state in which the plating can adhere to it easily.

In fields other than molds, this type of plating is used for preventing rust in the parts for food production machines, mill rolls, and for preventing rusting of machine parts.

In general, since the waste plating liquid treatment and management of plating liquid are important from the point of view of pollution prevention, in most cases, the work of hard chromium plating is entrusted to specialized companies.

캐비티의 표면 도금은 산화 또는 부식을 방지하기 위한 수단으로 사용된다. 경질 크롬 도금은 다양한 종류의 도금 중에서 전형적인 방법이다.

경질 크롬 도금은 내마모성 및 내식성이 우수하다.

표면의 피막의 경도는 700 ~ 1000 Hv이며, 표면은 흰색의 금속 광택이 있다.

도금을 수행하기 위해서 도금 될 작업 물을 무수 크롬산, 황산 등으로 이루어진 도금 용액에 담그고 전류를 통과시켜 전기 도금 막을 형성한다.

도금된 피막은 공기 중에서 비금속의 강력한 산화 피막이다. 도금된 막의 표면 에너지는 낮기 때문에 다른 재료에 의해 접착되기 어려운 성질을 갖는다. 따라서 플라스틱 성형품이 금형으로부터 이탈되기 쉬운 성질이 있다고 할 수 있다.

도금을 능숙하게 하기 위해서 도금할 표면을 세심히 연마하여 도금이 부착하기 쉬운 상태로 만드는 것이 바람직하다.

금형 이외의 분야에는 식품 생산 기계, 회전 롤용 부품의 녹 방지, 기계 부품의 녹 방지용으로 사용된다.

일반적으로 폐액의 처리 및 도금액의 관리는 오염 방지의 관점에서 중요하기 때문에 대부분의 경우 경질 크롬 도금의 작업은 전문 회사에 위탁된다.

H 012 Accuracy of Mold Preparation

The accuracy of preparation of parts of a mold for plastic injection molding is determined by the dimensional tolerance of the molded product. While each company may have its own internal standards about the policies for this accuracy, one of the standards used in major advanced countries is the standard recommended by VDI (Association of German Engineers).

The details of the molded product tolerances according to the VDI standard are as follows.

Taking the dimensional tolerance of the molded product as 100,
Mold preparation tolerance = 1/3 = 33%
Wear margin of mold = 1/6 = 17%
Fluctuation of molding shrinkage = 1/2 = 50%

Let us obtain the accuracy of mold preparation based on the above standards.

[Case 1]

What is the accu racy of mold preparation when the dimensional tolerance of the molded product is ±0.3 ㎜?

▶ [Example of investigations]

The dimensional tolerance of the molded product is +0.3+|-0.3|= 0.6 ㎜

(| | indicates the absolute value.)

H-012 금형 준비의 정밀도

사출성형을 위한 금형 부품의 정밀도는 성형 제품의 치수 공차에 의해 결정된다. 각 회사마다 이 정밀도에 대한 정책에 대한 내부 표준이 있을 수 있지만 주요 선진국에서 사용되는 표준 중의 하나는 VDI (독일 엔지니어 협회)에서 권장하는 표준이다.

VDI 표준에 따른 성형품 공차의 세부 사항은 다음과 같다.

성형품의 치수 공차를 100,
금형의 공차 = 1/3 = 33%
금형의 마모 = 1/6 = 17%
성형 수축에 의한 변동 = 1/2 = 50%

위의 표준을 기반으로 금형의 정밀도를 확보하자.

[사례 1]

성형품의 치수 공차가 ±0.3㎜인 경우, 금형 제작의 정밀도는 어떻게 될까?

▶ [조사 사례]

성형품의 치수 공차는 +0.3 + | -0.3 | = 0.6 ㎜

(| |는 절대 값을 나타낸다.)

Therefore, the tolerance of mold preparation is 0.6 × 1/3 = 0.2 ㎜.

∴ A mold preparation accuracy of ±0.1 ㎜ is appropriate.

따라서 금형의 공차는 0.6 × 1/3 = 0.2 ㎜ 이다.

∴ 금형의 정밀도는 ± 0.1 ㎜가 적당하다.

[Case 2]

What is the accuracy of mold preparation when the dimensional tolerance of the molded product is ±0.05 ㎜?

[사례 2]

성형품의 치수 공차가 ± 0.05 ㎜인 경우 금형 제작의 정밀도는 어떻게 될까?

▶ [Example of investigations]

The dimensional tolerance of the molded product is +0.05+|-0.05|= 0.1 ㎜

Therefore, the tolerance of mold preparation is 0.1 × 1/3 = 0.03 ㎜.

∴ A mold preparation accuracy of ±0.015 ㎜ is appropriate.

▶ [조사 사례]

성형품의 치수 공차는 +0.05 + | -0.05 | = 0.1 ㎜

따라서 금형 공차는 0.1 × 1/3 = 0.03 ㎜ 이다.

∴ 금형 정밀도는 ± 0.015 ㎜가 적당하다.

H 013 Thermal Stress on Hot Runner Manifold

In a hot runner mold, usually a manifold is used for diverting the molten plastic resin injected from the nozzle of the injection molding machine to the hot runner. A manifold is a plate for branching to the runner, has a heater incorporated in it, and hence the plastic resin inside the manifold is maintained in the molten state and is maintained in a state in which it can flow.

H-013 핫 런너 매니 폴드의 열 스트레스

일반적으로 핫 런너 금형에서 매니 폴드는 사출성형기의 노즐로부터 주입 된 용융수지를 핫 런너로 전달하는데 사용된다. 매니 폴드는 런너가 분리된 판에 히터가 내장되어 있어 매니 폴드 내의 플라스틱 수지가 용융 상태로 유지되고, 유동 가능한 상태로 존재한다.

A large amount of heat will be accumulated in the manifold in order to maintain the plastic in the molten state, and as a result, the steel material itself of the manifold undergoes thermal expansion.

Since the fact that the steel material undergoes thermal expansion implies that its external dimensions expand, this in turn generates compression stresses. Such stress generated due to heat is called "thermal stress".

Unless we know the size of the thermal stress, a large compression stress will act on the mold, particularly on the mold plate on the fixed side and on the mold base components, which can lead to deformations of the cavity and core, burrs, and can also cause breakage of the cavity in extreme cases.

In view of this, in this lesson we learn the method of calculating the thermal stress acting on the manifold. The thermal stress σ (sigma) acting on the manifold can be calculated using the following equation.

$$\sigma = \varepsilon \cdot E = \Delta l \cdot E/L$$

Where,

ε (epsilon): strain (%), and

E: Young's modulus of the manifold steel material (MPa or kgf/㎟). In the case of carbon steel, E = 20,000 to 21,000 kgf/㎟.

Δl (delta) = Amount of thermal expansion of length (㎜)

L: Overall length of the manifold (㎜)

플라스틱을 용융 상태로 유지하기 위해 다량의 열이 축적되어 매니 폴드의 강재 자체가 열팽창을 하게 된다.

강재가 열팽창 한다는 사실은 외형 치수가 확장됨을 의미하므로 압축 응력을 발생시킨다. 이러한 열에 의한 응력을 열응력이라 한다.

열응력의 크기를 알지 못하면 금형, 특히 고정측 플레이트와 몰드베이스 구성 요소에 큰 압축 응력이 작용하여 캐비티와 코어의 변형, 버를 유발할 수 있다. 극단적인 경우에 캐비티가 파손될 수 있다.

이를 보면서 이 단원에는 매니 폴드에 작용하는 열응력을 계산하는 방법을 학습한다. 매니 폴드에 작용하는 열응력 σ는 다음 식을 사용하여 계산할 수 있다.

$$\sigma = \varepsilon \cdot E = \Delta l \cdot E/L$$

여기서,

ε : 변형률 (%)

E : 매니 폴드 강재의 영율 (MPa 또는 kgf/㎟). 탄소강의 경우, E = 20,000 내지 21,000 kgf/㎟.

Δl = 열팽창 총 길이 (㎜)

L : 매니 폴드의 전체 길이 (㎜)

Here, Δl can be calculated as follows.

$$\Delta l = \alpha \cdot \Delta t \cdot L$$

Where, α (alpha): Linear thermal expansion coefficient of the manifold steel material (㎜/℃)

In the case of carbon steel, α is about 11.5 to 12.8×10^{-6}(㎜/℃).

Δt: Temperature change from room temperature up to the set temperature of the manifold (℃)

The thermal stress can be obtained from these equations.

In the next lesson, we will practice some actual examples of calculations.

여기서, Δl은 다음과 같이 계산 될 수 있다.

$$\Delta l = \alpha \cdot \Delta t \cdot L$$

여기서, α : 매니 폴드 강재의 선형 열팽창 계수 (㎜/℃)

탄소강의 경우, α는 약 11.5 ~ 12.8 × 10^{-6} (㎜/℃).

Δt : 실내 온도에서 매니 폴드의 설정 온도까지의 온도 변화 (℃)

열 응력은 이 방정식에서 얻을 수 있다.

다음 단원에는 계산의 실제 예를 몇 가지 연습 할 것이다.

H 014 Thermal Stress on Hot Runner Manifold (Practice Calculations)

H-014 핫 런너 매니 폴드의 열 스트레스(실습 계산)

Question

Calculate the thermal stress σ acting on the manifold in the following example of settings.

However, your calculations should be based on the assumptions given below.

Overall length of the manifold L = 150 ㎜

Room temperature of a molding factory in Japan in December = 18℃

Plastic used = ABS plastic

Manifold steel material = S55C

문제

다음 설정 예에서 매니 폴드에 작용하는 열응력 σ를 계산하자.

그러나 계산은 아래 주어진 가정을 기반으로 해야 한다.

매니 폴드의 전체 길이 L = 150 ㎜

12월의 일본 성형 공장의 룸 온도 = 18℃

Plastic = ABS 플라스틱

매니 폴드 스틸 재질 = S55C

▶ Sample Answer

Firstly, we obtain the change in the dimensions Δl due to temperature rise as follows.

$$\Delta l = \alpha \cdot \Delta t \cdot L$$

Where,

α (alpha): Linear thermal expansion coefficient of the manifold steel material (㎜/℃)

In the case of S55C, we assume that α is equal to 12×10^{-6}(㎜/℃)).

Δt: Temperature change from room temperature up to the set temperature of the manifold (℃)

$$\therefore \Delta l = \alpha \cdot \Delta t \cdot L = 12 \times 10^{-6} \times (230-18) \times 150 = 0.3816\ (mm).$$

Therefore, the thermal stress σ is given by:

$$\sigma = \varepsilon \cdot E = \Delta l \cdot E/L$$

Where,

ε (epsilon): strain (%), and

E: Young's modulus of the manifold steel material (MPa or kgf/㎟).

In the case of S55C, E = 21,000 kgf/㎟.

Δl (delta) = Amount of thermal expansion of length (㎜)

L: Overall length of the manifold (㎜)

▶ 답안 예

먼저 온도 상승에 의한 Δl의 변화를 다음과 같이 구한다.

$$\Delta l = \alpha \cdot \Delta t \cdot L$$

여기서,

α : 매니 폴드 강재의 선형 열팽창 계수 (㎜/℃)

S55C의 경우, α는 12×10^{-6} (㎜/℃).

Δt : 매니 폴드의 실내 온도에서 설정 온도까지의 온도 변화 (℃)

$$\therefore \Delta l = \alpha \cdot \Delta t \cdot L = 12 \times 10^{-6} \times (230-18) \times 150 = 0.3816\ (mm)\text{이다.}$$

그러므로 열응력 σ는 다음과 같이 주어진다.

$$\sigma = \varepsilon \cdot E = \Delta l \cdot E/L$$

여기서,

ε (ε) : 변형률 (%) 및

E : 매니 폴드 강재의 영율 (MPa 또는 kgf/㎟)

S55C의 경우, E = 21,000 kgf/㎟.

Δl = 열팽창 길이 (㎜)

L : 매니 폴드의 전체 길이 (㎜)

$$\therefore \sigma = \varepsilon \cdot E = \Delta l \cdot E/L$$
$$= 0.3816 \times 21{,}000/150$$
$$= 53.424\,(kgf/mm^2)$$
$$= 5342.4\,(kgf/cm^2)$$

$$\therefore \sigma = \varepsilon \cdot E = \Delta l \cdot E/L$$
$$= 0.3816 \times 21{,}000/150$$
$$= 53.424\,(kgf/mm^2)$$
$$= 5342.4\,(kgf/cm^2)$$

H 015 Removable insert

In the design of molded article, quite often there will be a need for undercut shapes. (An undercut is a shape that cannot be taken out just by opening the parting surface. For example, a horizontal hole, a projection, a depression, etc.).

In the case of an undercut shape, it will be necessary to design the mold using mechanical engineering techniques so that the molded article can be taken out with a special mold structure such as a slide core or a slant pin, etc.

The design would be easy if the undercut can be handled using the simplest slide core structure, but sufficient knowledge of and expertise in mold design will be required if a complex undercut shape has to be produced.

When a mounting test is to be done by quickly obtaining a sample of the molded article with a trial mold, etc., the delivery time can become as long as two weeks if the development of a special undercut mechanism and mold preparation is to be made. In such

H-015 착탈식 인서트

성형품의 디자인에 있어서 가끔 언더컷 형태가 필요하다. (언더컷은 파팅면을 열어서 추출 할 수 없는 모양이다. 예를 들어 수평 구멍, 돌출부, 함몰 등).

언더컷 형상의 경우에는 슬라이드 코어 또는 경사 핀과 같은 특수 금형 구조로 취출 될 수 있도록 기계적 엔지니어링 기술을 이용하여 금형설계를 할 필요가 있다.

가장 간단한 슬라이드 코어 구조를 사용하여 언더컷을 처리 할 수 있다면 설계가 쉽지만 복잡한 언더컷 형상을 제작해야하는 경우 금형 설계에 대한 충분한 전문 지식이 필요하다.

시험 금형 등으로 성형품의 샘플을 신속하게 확보하기 위하여 장착 시험을 행하는 경우 특별한 언더컷 메커니즘과 금형 준비가 이루어지려면 납기가 2주 정도 길어질 수 있다.

situations, a method of handling undercuts called the "removable insert" is used which was devised by our seniors with some skillful techniques.

What is a "removable insert"?

- The core and the cavity that constitute the part of the molded article that becomes an undercut are divided, and then designed and fabricated.
- The removable inserts that are prepared by dividing the mold are taken out from the mold along with the molded product at the time of ejecting the molded article.
- The removable inscrt is removed from the molded article by a manual operation.
- The removable insert is installed in the mold manually at the time of the next molding.
- In other words, with a removable insert structure, it is not possible to carry out continuous injection molding, and the method is that of forming the undercut shape semi-automatically using manual operations additionally.
- There are also some cases in which the removable insert is pushed out from the mold by an ejector pin.
- In order to position the removable insert, a key structure or a knock pin structure is used in some ca3c3.
- There are cases in which the molding is done after pre-heating the removable insert.
- There is also the method of molding which makes the semi-automatic molding more efficient by preparing two removable inserts, and inserting them alternately into the mold.

이러한 상황에서 우리의 선배가 숙련된 기술로 고안한 "착탈식 인서트"라는 언더컷 처리 방법이 사용된다.

"착탈식 인서트"란 무엇인가?

- 언더컷이 되는 성형품 부분을 구성하는 코어 및 캐비티를 분할하여 설계 제작한다.
- 금형을 분할하여 준비되는 착탈식 인서트는 성형품을 취출할 때 금형과 함께 취출된다.
- 제거 가능한 인서트는 수동 조작으로 성형품에서 제거된다.
- 세거 가능한 인서트는 다음 성형 할 때 수동으로 금형에 삽입된다.
- 즉, 착탈식 인서트 구조로 연속 사출성형을 수행 할 수 없으며, 수동 조작을 추가로 적용하여 반자동 형상의 언더컷을 수동 조작으로 성형하는 방법이다.
- 탈착식 인서트가 이젝터 핀에 의해 금형으로부터 밀려 나오는 경우도 있다.
- 착탈식 인서트의 위치를 결정하기 위해 키 구조 또는 노크 핀 구조가 사용되는 경우가 있다.
- 착탈식 인서트를 예열한 후에 성형이 이루어지는 경우가 있다.
- 또한 2개의 탈착식 인서트를 준비하고, 이를 교대로 금형에 삽입하여 반자동 성형을 보다 효율적으로 수행 할 수 있는 성형 방법이 있다.

- In order to ensure that the direction of inserting the removable insert is not mistaken, there is also the technique of using a direction determining structure.
- The key point is to design the escape and the clearance so as to make it easy to install and remove the removable inserts.
- There is also the management aspect of provisionally preparing the trial mold with a removable insert structure, thereby obtaining time for developing a mold structure such as the slide core structure that can permit continuous molding.

- 착탈식 인서트 삽입 방향을 잘못 이해하지 않도록 방향 결정 구조를 사용하는 기술도 있다.
- 핵심은 착탈 가능한 인서트를 쉽게 설치하고, 제거 할 수 있도록 탈출구와 간극을 설계하는 것이다.
- 착탈식 인서트 구조로 시험 금형을 임시적으로 준비하는 경우도 있으므로 연속 성형이 가능한 슬라이드 코어 구조와 같은 몰드 구조를 개발할 수 있는 시간을 확보 할 수 있다.

I. 유지보수

CHAPTER I

I 001 Why Does a Mold Break?

Although no unnecessary force is applied to a mold when it is being assembled, when it is actually installed in an injection molding machine and the molding operation being conducted, it is subject to various external forces unlike those experienced during assembly.

For example, the mold clamping force when a mold is being clamped can be from several tons to several hundreds of tons, even several thousands of tons. It is necessary that the mold has enough strength to withstand that compression stress.

In addition, in order to completely fill the mold with molten plastic via the sprue and via the runner, it is necessary to apply pressure to the plastic and make it flow inside the mold. The reason for this is that since molten plastic is a fluid having viscosity, a sufficient pushing force is necessary to make it flow into the mold.

The force of the pressure can be 1000 to 2000 kgf/cm^2 near the sprue inlet, and even inside the cavity the force of the pressure is 200 to 600 kgf/cm^2.

In addition, since the force of the pressure acts for a very short time which is normally not even 1 second, considerable shock is applied to the core pin and the walls of the cavity, and in some cases, this may cause buckling of the thin and long pin.

I-001 금형 고장의 원인은?

금형을 조립할 때 금형에 불필요한 힘이 가해지지는 않지만 사출성형기에 실제로 설치되어 성형 작업이 진행될 때에는 조립 시와 달리 다양한 외력이 가해진다.

예를 들어 금형을 클램핑 할 때의 형체력은 수 톤에서 수백 톤, 심지어 수천 톤이 될 수 있다. 금형은 압축 응력을 견딜 수 있는 강도를 가져야한다.

또한 금형을 스프루와 런너를 통해 용융수지로 완전히 채우기 위해서 플라스틱에 압력을 가하고 금형 내부로 흐르게 할 필요가 있다. 그 이유는 용융수지가 점성을 갖는 유체이기 때문에 금형 내로 유동시키기에 충분한 압력이 필요하기 때문이다.

압력은 스프루 입구 근처에서 1000 ~ 2000 kgf/cm^2가 될 수 있으며, 캐비티 내부에서도 200 ~ 600 kgf/cm^2의 압력이 가해진다.

또한, 압력은 보통 1 초도 되지 않는 매우 짧은 시간 동안 작용하기 때문에 코어 핀 및 캐비티의 벽면세 상당한 충격이 가해지며, 경우에 따라서는 얇고 긴 핀이 좌굴 될 수 있다.

In this way, if we sequentially look at the process by which the parts of a mold break, we can find the corresponding causes. In order to make sure that a mold does not break, at the time of designing a mold, it is very important to make clear the basic environment of use (injection pressure, mold structure, acting stresses, etc.) in terms of numerical values, and to verify in advance the actual operation of the mold. This is because fatal damages can occur that cannot be covered by fine adjustments after the mold has been prepared if the mold preparation is done without carrying out the strength calculations of the basic structure and while defects are allowed to be present in the structure.

In addition, even when machining the parts of a mold or at the time of assembling and adjustment, it is very important to give considerations to machining after understanding the shapes of the parts, the surface quality, the accuracy of mating, etc.

In the case of machining, although the minimum possible responsibilities can be said to have been carried out as long as the work has satisfied the dimensions, accuracy, and tolerances specified in the drawings, in order to make a more superior mold, it is desirable to understand the functions of all the parts of the mold, so as to advance one step further.

In order to prepare molds that do not break, it is very important that there is a balance between the basic concepts and the considerations in machining and assembly.

이런 방식으로 금형의 부품이 부서지는 과정을 순차적으로 살펴보면 해당 원인을 찾을 수 있다. 금형설계 시 금형이 파손되지 않도록 하기 위해서 수치적으로 사용되는 기본 환경 (사출압력, 금형 구조, 작용 응력 등)을 명확히 하고, 금형의 실제 작동을 미리 확인하는 것이 매우 중요하다.

왜냐하면 기본적인 구조 강도 계산을 수행하지 않고 금형을 제작하고, 구조적 결함이 포함된 금형은 미세 조정으로 해결할 수 없는 치명적인 손상이 발생할 수 있기 때문이다.

또한, 금형 부품을 가공하거나 조립 및 조정할 때에도 부품의 형상, 표면 품질, 결합 등의 정확도를 이해한 후 가공에 대한 고려가 이루어져야 한다.

기계 가공의 경우, 도면에 명시된 치수, 정밀도 및 공차를 만족한다면 최소한의 책임은 다했다고 할 수 있지만, 보다 우수한 금형을 만들기 위해서 금형의 모든 부분의 기능을 이해해야만 한 단계 더 나아갈 수 있다.

파손되지 않는 금형을 준비하려면 가공과 조립 시 기본 개념과 고려 사항들 사이에서 균형이 있어야 한다.

I 002 Periods of Mold Failures

Machines and manufacturing facilities that contain molds may have their functions stopped or seriously curtailed due to "failure". It is known that failures can be classified statistically into three classes depending on the period of their occurrence.

Failures are of the following three types.

	I	II	III
Period name	Initial failure period (initial failure)	Random failure period (random failure)	Wear-out failure period (wear-out failure)
Main causes	Production quality defect	Design quality defect	Life
Remarks		Difficult to predict the time of failure occurrence	It is possible to estimate the period of wear-out Countermeasures can be taken by preventive maintenance

1. Initial failure period (Initial failure)

This is the period in which failures occur due to quality defects in the components of molds, etc., or due to basic design mistakes. Since such failures occur soon after starting the use of molds, etc., it will not be possible to start continuous production.

I-002 금형 고장 기간

금형을 포함한 기계 및 제조 설비는 "고장"으로 인해 기능이 중지되거나 심각하게 축소 될 수 있다. 고장은 발생 기간에 따라 통계적으로 세 가지 등급으로 분류 될 수 있다.

고장은 다음의 세 가지 유형이 있다.

	I	II	III
기간 이름	초기 고장 기간 (초기 불량)	임의의 고장 기간 (임의의 실패)	마모 고장 기간 (마모 고장)
주요 원인	생산 품질 불량	설계 품질 불량	수명
비고		장애 발생 시간을 예측하기 어려움	마모 고장에 의한 기산 계산이 가능. 예방 정비로 유지가능

1. 초기 고장 기간 (초기 고장)

이것은 금형 등의 구성 요소의 품질 결함 또는 기본 설계 실수로 인해 오류가 발생하는 기간이다. 이러한 고장은 금형 사용을 시작한 후 곧 발생하기 때문에 연속 생산을 시작할 수 없다.

2. Random failure period (random failure)

This is the period in which failures occur due to some cause in the middle of stable operations that are being made for some time using molds, etc., for continuous production.
The causes of such failures can be insufficient quality level in the design, which can cause some of the components to exhaust their lives, or due to some component breaking due to the effect of more than expected stress, etc.

Since it is difficult to predict the period of occurrence of failures during this period, this period is being referred to as the random failure period.

2. 임의의 고장 기간(임의의 고장)

이것은 연속 생산을 위해 금형 등을 사용하여 일정 기간 안정된 작업이 진행되는 동안 어떠한 원인에 의해 고장이 발생하는 기간이다.

일부 구성 요소가 수명을 다하거나 예상보다 많은 스트레스 등으로 인해 구성 부품이 파손되는 원인은 수준 미달의 설계가 원인일 수 있다.

이 기간 동안 고장 발생 기간을 예측하기 어렵기 때문에 이 기간을 무작위 고장 기간이라고 한다.

3. Wear-out failure period (wear-out failure)

This is the period in which failures occur during continuous production using molds, etc., due to their being used near the expected life in terms of the number of shots.
The cause of such failures is the wearing out of components. The period when wear-out reaches the limit of usability can be predicted using sampling and by experience.

Therefore, it can be said that it is possible to some extent to avoid the occurrence of wear-out failures by preventive maintenance.

The relationship between the failure period and the failure rate plotted as a curve as shown in the figure is called the bath tub curve because the curve of failure rate with respect to the time axis is similar to the cross-sectional shape of a bath tub.

3. 마모 고장 기간 (마모 고장)

이것은 금형을 사용하여 연속 생산하는 동안 고장이 발생하는 기간으로 사출 횟수와 관련하여 예상 수명 근처에 도달하기 때문에 발생한다. 이러한 고장의 원인은 부품의 마모 때문이다. 마모가 사용 한계에 도달하는 기간은 샘플링과 경험을 통해 예측할 수 있다.

따라서 예방 유지 보수를 통해 마모 고장이 발생하지 않도록 하는 것이 어느 정도까지는 가능하다고 할 수 있다.

도면에 도시된 바와 같이 고장 시간과 고장율과의 관계는 시간축에 대한 고장율 곡선이 욕조의 단면형상과 유사하기 때문에 bath tub curve라 불린다.

Fig. The bath tub curve of failure rate

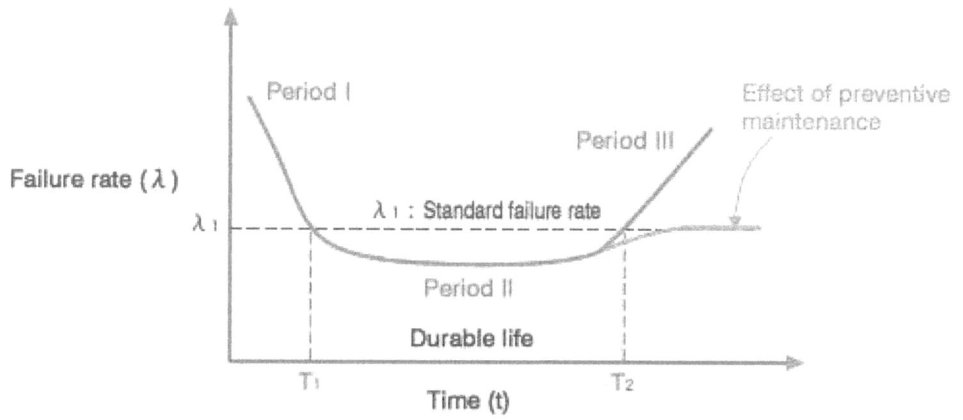

I 003 Mean Time Between Failures of Molds (MTBF)

In plastic injection molding, when molding operations are being carried out in mass manufacturing, initial failures or random failures can occur at certain probabilities. In the last course we explained the different failure periods, and this time we will explain what is called the "Mean Time Between Failures (MTBF)".

The average time after a mold has failed until the next failure occurs is called the MTBF. Therefore, the unit of MTBF used is time (hr, min, etc.) or the number of shots.

When the value of MTBF is large, it can be evaluated that the mold is one in which failure is difficult to occur.

I-003 평균 금형 고장 시간 (MTBF)

사출성형에서 대량 생산 시 성형 작업을 수행하는 경우 초기 고장 확률 또는 임의 고장 확률이 특정 확률로 발생할 수 있다. 마지막 코스에는 다양한 고장 기간에 대해 설명했으며, 이번에는 MTBF(Mean Time Between Failures)를 설명한다.

금형이 고장 나서 다음 고장이 발생할 때까지의 평균시간을 MTBF 라고 한다. 따라서 사용 된 MTBF 의 단위는시간 (시간, 분 등) 또는 사출 횟수이다.

MTBF의 값이 클 때, 금형은 고장이 발생하기 어려운 금형이라고 평가 할 수 있다.

On the other hand, if the value of MTBF is small, it can be said that the mold is likely to fail easily, and it is difficult to make a plan of production of stable molding operations.

Let us consider MTBF in a more realistic example. When several repeated molds are manufactured using the same design drawings, although initial failures and random failures are more likely to occur in the first mold, if technical improvements are made in response to such failures, it will be difficult for such failures to occur in the second and subsequent molds. Therefore, the MTBF of second and subsequent molds will be a large number compared to that of the first mold.

When considering the "production cost + maintenance and management cost" of a mold from it is initially produced until it is disposed off, if the cost of a mold is evaluated based only on the fact that the initial cost (initial production cost) of the mold is cheap, in the end it may become an expensive purchase.

MTBF has a lot of importance as an index for evaluating the maintenance and management cost.

On the other hand, as an index for evaluating the ease of repair of molds, there is what is called the Mean Time To Repair (MTTR).

In case a mold fails, the MTTR of that mold has a small value if it is possible to repair it quickly.

As an example, if spare parts are always in stock, and also if the structure is such that it is possible to replace the spare parts from the parting surface even without removing the mold plate of the cavity, then the MTTR will become a small value.

MTTR, also, has a lot of importance as an index for evaluating the maintenance and management cost.

반면, MTBF의 값이 작으면 금형이 쉽게 파손될 우려가 있어 안정된 성형 작업의 계획을 세우기 어렵다.

보다 현실적인 예를 들어 MTBF를 고려해 보겠다. 동일한 설계 도면을 사용하여 여러 개의 반복 된 금형을 제조하는 경우 첫 번째 금형에서 초기 불량 및 무작위 고장이 발생할 가능성이 있지만 이러한 불량에 대한 기술적 개선이 이루어지면 두 번째 및 후속 금형에는 해당 금형에서 발생하는 오류가 발생하기 어렵다. 따라서 두 번째 및 후속 금형의 MTBF는 첫 번째 금형의 MTBF에 비해 크다.

금형의 "생산 원가 + 유지 관리 및 관리 비용"을 금형이 폐기 될 때까지 고려할 때, 금형의 초기 비용(초기 생산 비용)이 싸다는 사실만을 근거로 금형의 비용을 평가한다면, 결국에는 값 비싼 구매가 될 수 있다.

MTBF는 유지 보수 및 관리 비용을 평가하는 지표로서 중요하다.

한편, 금형의 보수의 용이성을 평가하는 지표로서, 평균 금형의 수리시간(MTTR)이 있다. 금형이 고장날 경우에 신속하게 복구 할 수 있다면, 그 금형의 MTTR는 작은 값을 갖는다. 예를 들어, 예비 부품의 재고가 항상 있을 경우, 또한 캐비티 형판을 제거하지 않고 파팅면에서 부품을 교체 할 수 있는 구조일 경우, MTTR은 작은 값이 된다.

MTTR은 또한, 유지 및 관리 비용을 평가하는 지표로서 중요하다.

MTBF and MTTR

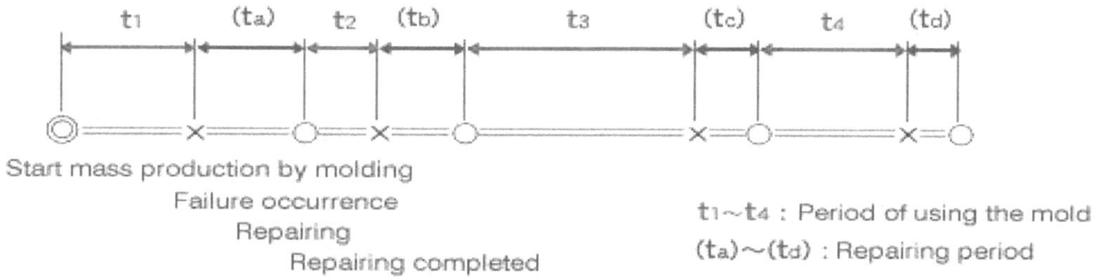

$$MTBF = \frac{\sum_{n=1}^{n} tn}{n} = \frac{t1+t2+t3+t4}{4}$$

(Mean Time Between Failures)

$$MTTR = \frac{\sum_{n=a}^{n} tn}{n} = \frac{ta+tb+tc+td}{4}$$

(Mean Time To Repair)

I 004 Key Points of Maintenance of Molds

When molds for plastic injection molding are used for molding in mass production, the performance of molds decreases gradually due to wear, rust, etc. Maintenance (maintenance management) is necessary in order to repair such decreases in performance.

It is recommended to pay attention to the following points as the key points in the maintenance of molds for plastic injection molding.

I-004 금형의 유지 관리의 요점

사출금형을 대량 생산용으로 사용하면 마모, 녹 등으로 인해 금형 성능이 서서히 저하된다. 이러한 성능 저하를 보완하기 위해서 유지 보수 (유지 관리)가 필요하다.

사출금형의 유지 관리 요점은 다음과 같다.

1) Scratches in the surface of the core pin
2) Wear in the surface of the core pin
3) Fusing of the surface of the core pin
4) Depression in the butting surface of the core pin
5) Scratches and corrosion in the mirror finished parts
6) Peeling off of electroplating
7) Wear and deformation of the submarine gate hole
8) Wear and deformation of the pin point gate hole
9) Depression in the periphery of the parting surface
10) Wear of the ejector pin hole
11) Biting of the ejector sleeve hole
12) Soot collected in the air vent parts
13) Biting of the sliding surface of the slide core
14) Weakening of the coil springs
15) Warping and deformation of the cavity frame
16) Cracks in the corner parts of the cavity
17) Biting of the ejector guide pin and bush
18) Wear in the nozzle touching surface of the sprue bush
19) Adhesion of scale and rust in the cooling water hole
20) Water leakage in the cooling water hole

1) 코어 핀의 표면 스크래치
2) 코어 핀의 표면 마모
3) 코어 핀의 표면 용융
4) 코어 핀의 접촉 표면 함몰
5) 경면사상 부품의 흠집과 부식
6) 전기 도금 박리
7) 서브마린 게이트의 마모 및 변형
8) 핀 포인트 게이트의 마모 및 변형
9) 파팅면 주변에서의 움푹 패임
10) 이젝터 핀 구멍의 마모
11) 이젝터 슬리브 구멍의 뜯김
12) 공기 벤트 부품에 쌓인 그을음
13) 슬라이드 코어의 슬라이딩면 뜯김
14) 코일 스프링의 약화
15) 캐비티 프레임의 뒤틀림 및 변형
16) 캐비티 코너 부분의 균열
17) 이젝터 가이드 핀 및 부시의 뜯김
18) 스프루 부시의 노즐 접촉면에서의 마모
19) 냉각수 구멍의 스케일 및 녹의 부착
20) 냉각수 구멍의 누수

I 005 Techniques of Mold Maintenance

After molding is carried out, it is necessary to carry out maintenance of the plastic injection molds such as disassembling and cleaning, etc. (maintenance and inspection work).

I-005 금형 유지보수 기술

성형 작업 후에는 분해 및 세정 등의 금형 유지보수(보수 점검 작업)가 필요하다.

The "soot" constituents generated from the molding material, and the condensed gas deposits, etc. accumulate on the surface of the cavity or core and on the parting surface.

Although these constituents are still in the liquid state when the mold is still at a high temperature, but when the mold cools down they solidify, moisture content in air gets adhered because of these, eventually causing the generation of rust and corrosion of the mold.

When this state is reached, the surface quality of the molded item decreases, there will be fluctuations in the dimensional accuracy, insufficient air escape, which causes the generation of short shots.

The frequency of maintenance varies widely depending on the status of quality management of the molded item, and on the size of the mold, etc.

In the case of molds with short periods, it is necessary to carry out mold maintenance once in a few days, and about once in two months in the case of mold with long periods.

A common method of maintenance is to disassemble the mold, cleaning each of the part by ultrasonic cleaning, or by rinsing in an organic solvent, and to remove the rusted part by polishing and then electroplating.

In the case of parts having a lot of rust, the parts are replaced, or the inserting part is corrected.

It is necessary to modify tools, jigs, and equipment in order to carry out maintenance efficiently.

Cleaning brush, disassembling tools, wooden tools, cloth jigs, buffing tools and sand paper, polishing cream, lapping material, special chop sticks, bamboo spatula, bamboo comb, air tools, crane supplementary tools, Z light, magnifying glass, etc., need to be modified.

성형 재료로부터 생성된 "그을음 "성분 및 응축 된 가스 침전물 등은 캐비티 또는 코어의 표면 및 파팅면상에 축적된다.

금형이 고온일 때에 이들 요소는 여전히 액체 상태이지만 금형이 냉각되면 고화되고, 공기 중의 수분 이 달라붙어 결과적으로 녹의 발생과 금형 부식을 유발하다.

이 상태에 도달하면 성형품의 표면 품질이 하락하며 치수 정밀도가 불균일해지고, 에어벤트가 불충분하여 미성형이 발생한다.

유지 관리 빈도는 성형품의 품질 관리 상태 및 금형 크기 등에 따라 크게 다르다.

짧은 사이클 타임의 금형인 경우, 금형 유지 관리를 며칠에 한 번씩, 긴 사이클 타임의 금형의 경우 약 2개월에 한 번 수행해야 한다.

유지 보수의 일반적인 방법은 금형을 분해하고 각 부품을 초음파 세척 또는 유기 용제로 헹구는 것, 연마 한 후 전기 도금으로 녹슨 부품을 제거하는 것이다.

녹이 많은 부품의 경우 부품을 교체하거나 삽입 부품을 수정한다.

유지 보수를 효율적으로 수행하려면 도구, 지그 및 장비를 수정해야한다

청소 브러쉬, 분해 도구, 나무 도구, 헝겊 지그, 버핑 도구 및 샌드페이퍼, 연마 재료, 랩핑 재료, 특수 젓가락, 대나무 주걱, 대나무 빗, 공기 도구, 크레인 보조 도구, Z 조명, 돋보기 등의 수정이 필요하다.

Even the work bench needs to be modified so that the work becomes easy. Modifications are to be done about the work bench height to improve ease of carrying out the work, walking space, crane position, air gun piping, etc.

심지어 작업 벤치를 조절하여 작업을 쉽게할 필요가 있다. 작업 수행의 용이함, 작업 공간, 크레인 위치, 에어컨 파이프 등이 개선된다.

It is also useful to use a table or white board for spreading the drawings, a digital camera, and a video camera.

또한 도면을 펼칠 테이블이나 화이트 보드, 디지털 카메라 및 비디오 카메라를 사용하는 것이 유용하다.

In order to increase the efficiency of maintenance, it is necessary to give considerations from the point of designing the mold.

유지 보수의 효율성을 높이려면 금형설계 시점부터 고려되어야 한다.

Holes are provided so as to facilitate disassembly of the mold, screws for hook bolts are provided, and supplementary screw holes are provided for assembling.

금형의 분해를 용이하게 하기 위한 구멍을 만들고, 후크 볼트용 나사부를 가공하며, 조립을 위한 보조 나사 구멍이 필요하다.

The cores should have the frame block construction so that they do not come apart, and it is also effective to use a construction of fixing using keys.

코어는 분리되지 않도록 프레임 블록 구조를 만들고, 키를 사용하여 고정 구조를 사용하는 것도 효과적이다.

Very often the placement numbers of core pins are engraved, and tag cuts are made in order to fix the directionality.

코어 핀의 배치 번호를 새기고, 꼬리표(번호표)를 만들어 조립 순서를 표시한다.

I 006 Analysis of Mold Breaking Accident

I-006 금형 사고 분석

When molding is being carried out using a mold, sometimes accidents occur in which a part gets damaged or broken due to some reason. Since a mold is a machine, the parts constituting a machine can get damaged due to some reason. The important thing is to find out the cause of such damage, and to take technical measures to prevent such an accident from occurring again.

금형을 사용하여 성형 할 때 어떤 이유로 어떤 부품이 손상되거나 파손되는 사고가 발생한다. 금형은 기계이기 때문에 어떤 이유로 기계를 구성하는 부품이 손상 될 수 있다. 중요한 것은 그러한 손상의 원인을 찾아내는 것이며 그러한 사고가 다시 발생하지 않도록 기술적 조치를 취하는 것이다.

If repairs are done while the reason for the damage is still vague and no fundamental technical measures have been taken, the same accident may occur again, or can even have the danger of causing harm to the workers.

Analysis of an accident of mold damage starts with detailed investigative analysis.

손상의 원인이 아직 모호하고 근본적인 기술적 조치가 취해지지 않은 상태에서 수리가 이루어지면 동일한 사고가 다시 발생할 수 있으며 근로자에게 해를 끼칠 위험이 있다.

금형의 손상 분석은 상세한 조사 분석으로 시작된다.

Types of damages

Damage	Wear	Corrosion	Fatigue
Impact (shock)		Excessive stress	
Fracture	Breakage, rupture		

손상의 종류

파손	마모	부식	피로
충격		과도한 스트레스	
골절	파손, 파열		

Items to be investigated in the event of a mold damage accident.

(1) Investigation of materials

- Investigation of manufacturing history
- Composition analysis(chemical analysis, impurity analysis, etc.)
- Investigation of mechanical properties (tensile test, compression test, buckling test, impact test, hardness test, etc.)
- Fatigue
- Surface observation and evaluation (macro, micro)
- Investigation of surface conditions, investigation of residual stress
- Investigation by observation of fracture surface

금형 손상 사고 시 조사할 항목.

(1) 재료 조사

- 제조 이력 조사
- 성분 분석 (화학 분석, 불순물 분석 등)
- 기계적 특성 조사(인장 시험, 압축 시험, 좌굴 시험, 충격 시험, 경도 시험 등)
- 피로 회복
- 표면 관찰 및 평가 (매크로, 마이크로)
- 표면 상태 조사, 잔류 응력 조사
- 파면 관찰에 의한 조사

(2) Investigation of design standards

- Verification of the design presumptions

(2) 설계 기준 조사

- 설계 추정의 검증

(3) Investigation of usage environment

- Investigation of external operating forces (size, repetition, impact, creep, etc.)
- Investigation of external environment (temperature, humidity, chemical atmosphere)
- Investigation of status of repairs

(3) 사용 환경 조사

- 외부 작동력 (크기, 반복, 충격, 크리프 등)에 대한 조사
- 외부 환경 (온도, 습도, 화학적 환경) 조사
- 수리 상태 조사

(4) Simulation

- Investigations of material mechanics
- Verification using simulation software
- Reproduction experiment

(4) 시뮬레이션

- 재료 역학 조사
- 시뮬레이션 소프트웨어를 사용한 검증
- 재생산 실험

(5) Overall investigation

(5) 전반적인 조사

I 007 Mold Deposits

I-007 금형 침전물

Mold deposits are materials that are deposited in the gaps between the inserts of molds for plastic injection molding or in the corners of the cavity. When the deposits accumulate, defects occur such as short shot of the molded item, or the shape is not accurately transferred to the molded article.

금형의 침전물은 금형 내 인서트 부품 또는 캐비티 모서리 사이의 틈에 퇴적되는 물질이다. 침전물이 쌓이면 미성형이나 형상이 정확하지 않은 성형품이 성형되는 등의 결함이 발생한다.

The causes of deposits can be the deposition of chemical constituent in the plastic or can be due to the effect of the chemicals or oil traces on the surface of the mold. The moisture in air can get adhered to the gap between the inserts and can cause the generation of rust, deposits get accumulated on such rusted parts, and the accumulation of deposits grows.

Although it is difficult for the "soot component" or the gas component in the volatile gases generated from the plastic to get adhered when the mold temperature is high due to molding operation, if the mold is removed from the molding machine and the mold temperature decreases, the gas components, etc. become solidified and become deposits.

The mold will last long if the deposits are removed periodically by dismantling and cleaning the mold. At the time of designing the mold, it is important to consider the optimum periodicity at which to dismantle and clean the mold.

In addition, it is also effective to adopt a mold structure which makes it difficult for deposits to get accumulated. For example, methods such as providing a gas vent surrounding the core pin and providing a groove surrounding the ejector pin and the runner lock pin for gas to escape are used in practice. It is also effective to use rust resistant steel such as stainless steel for the material of the mold components, or to provide hard chromium plating or a PVD film coating.

There are also some cases in which the gaseous components are forcibly exhausted using a vacuum suction device, but this makes the structure of the mold complicated.

During maintenance of the mold, the adhered matter is removed carefully by ultrasonic cleaning or using organic solvents to degrease the mold.

침전물의 원인은 플라스틱의 화학적 성분이거나 화학물질의 영향 또는 금형 표면의 오일 등이 될 수 있다. 공기 중의 수분은 인서트 사이의 간극에서 녹의 발생을 초래할 수 있고, 녹이 많은 부품에 침전물이 축적되고, 그 축적이 증가 할 수 있다.

성형 작업으로 인해 금형 온도가 높을 때에는 플라스틱으로부터 생성된 휘발성 가스의 "매연 성분" 또는 가스 성분이 부착되기 어렵지만 성형기에서 금형을 제거하고 금형 온도가 낮아지면 가스 성분 등이 응고되어 침전물이 된다.

금형을 분해하고 청소하여 주기적으로 제거하면 수명이 길어진다. 금형을 설계 할 때 금형을 분해하고 청소할 최적의 주기를 고려하는 것이 중요하다.

또한, 침전물이 쌓이는 것을 어렵게 하는 금형 구조를 적용하는 것도 효과적이다. 예를 들어, 코어 핀을 둘러싸는 가스 벤트를 제공하고, 이젝터 핀을 둘러싸는 홈 및 가스가 빠져 나가기 위한 런너 록 핀을 제공하는 것과 같은 방법이 실제로 사용된다. 금형 구성 요소의 재질에 스테인리스 스틸과 같은 녹 방지 강을 사용하거나 경질 크롬 도금 또는 PVD 필름 코팅을 제공하는 것도 효과적이다.

또한, 진공 흡착 장치를 사용하여 가스 성분을 강제적으로 배기하는 경우도 있지만, 이는 금형의 구조를 복잡하게 한다.

금형 보수 시에 초음파 세정 또는 유기 용매를 사용하여 부착 된 물질을 조심스럽게 제거하여 금형을 탈지한다.

I 008 Maintenance Items of Molds for Plastic Injection Molding

When the mass production of plastic injection molded products is being carried out, there will always be some wearing out or breakage of parts of the mold for plastic injection molding. In such situations, it is necessary to carry out maintenance by replacing parts or making repairs.

The common items that require maintenance are the following.

- Depressions, scratches and wear of the parting surface
- Chipping or depressions on the corners of the cavity
- Wear and scraping of the locking block
- Wear and cracks of the angular pin
- Wear and scraping of the guide post
- Wear and scraping of the guide bush
- Scraping of the rail guide by the slide core
- Scraping of the center rail
- Wear and deformation of the gate
- Wear of the internal surface of the sprue bush
- Depressions and deformation of the nozzle touching part of the sprue bush
- Weakening of the spring elasticity
- Wear of the ejector pin
- Wear of the hole of the ejector pin
- Wear and scraping of the positioning block
- Wear and scraping of the return pin

I-008 플라스틱 사출성형을 위한 금형의 유지 보수 항목

플라스틱 사출성형 제품의 대량 생산이 진행될 때 금형 부품의 마모 또는 파손이 항상 존재한다. 이러한 상황에는 부품을 교체하거나 수리하여 유지 보수를 수행해야 한다.

유지 보수가 필요한 공통 항목을 다음과 같다.

- 파팅면의 함몰, 스크래치 및 마모
- 구멍의 모서리에 치핑이나 함몰
- 록킹 블록의 마모 및 긁힘
- 앵귤러 핀의 마모 및 균열
- 가이드 핀의 마모 및 긁힘
- 가이드 부시의 마모 및 긁힘
- 슬라이드 코어에 의한 레일 가이드 스크래핑
- 센터 레일 스크래핑
- 게이트의 마모 및 변형
- 스프루 부시의 내부 표면 마모
- 스프루 부시의 노즐 접촉 부분의 파임 및 변형
- 스프링 탄력성 약화
- 이젝터 핀의 마모
- 이젝터 핀의 구멍 마모
- 위치 결정 블록의 마모 및 긁힘
- 리턴 핀의 마모 및 긁힘

- Scraping of the stripper plate
- Scraping of the runner stripper plate
- Elongation of the bolt
- Wear to the bolt threads
- Wear of the coupler for the cooling water
- Punctures and open circuits in the cartridge heater
- Rust and clogging inside the hole for the cooling water
- Rust and mold in the periphery of the mold base
- Open circuits in the electrical wires and cracks in the cable covering
- Fault in the contacts of the limit switches
- Clogging due to soot in the air vents
- Deformation due to temporal changes in the frame block

- 스트리퍼 플레이트의 긁힘
- 런너 스트리퍼 플레이트의 긁힘
- 볼트의 길이 신장
- 볼트 나사산의 마모
- 냉각수 커플러의 마모
- 카트리지 히터의 구멍 및 회로 단락
- 냉각수 구멍 내부의 녹 및 막힘
- 몰드베이스 주변의 녹 및 금형
- 전선의 단락 및 케이블 커버의 균열
- 리미트 스위치의 접점 고장
- 에어 벤트의 그을음에 의한 막힘
- 프레임 블록의 일시적 변화에 의한 변형

J. 기타

CHAPTER J

J 001 MuCell Process Latest Injection Molding Technology of USA

Various new technologies were exhibited in the largest plastics industry exhibition of the millennium "NPE2000" held in Chicago USA during the second half of June 2000. Among them, one of the most unique new technologies that attracted a lot of attention was that of the "MuCell ® Process".

This technology was invented in Massachusetts Institute of Technology, and at present a related venture company called Trexel Inc. has the basic patent and has entered into licensing agreements with eight injection molding machine manufactures in the world. A feature of this technology is that in a special mixing and kneading cylinder, nitrogen (N_2) gas or carbon dioxide (CO_2) is sealed in the supercritical state, injected into the mold in a manner similar to ordinary injection molding, filled, cooled, and the molded product is taken out.

Although no differences can be observed between the obtained injection molded product and an ordinary injection molded product, you will recognize the difference once you take the product in your hands. That is, the product will clearly be lighter (by about 12%) than an ordinary molded product. If you cut the product to know the reason for this difference, you will find that although the surface is solid, the interior has innumerable very fine bubbles generated in it and the material will have a sponge shape (with a honeycomb structure).

J-001 미국의 가장 최신 사출 성형 기술 MuCell

2000년 6월 하반기 미국 시카고에서 개최 된 밀레니엄 "NPE2000"의 가장 큰 플라스틱 산업 전시회에서 다양한 신기술이 전시되었다. 그 중 가장 주목받는 신기술 중 하나는 "MuCell ® 프로세스"이다.

그 기술은 매사추세츠 공과대학에서 발명하였으며, 현재 Trexel Inc.라는 벤처 회사가 기본 특허를 가지고 있고, 세계 8 개의 사출성형기 제조사와 라이센스 계약을 체결하고 있다. 이 기술의 특징은 특수 혼합 및 혼련 실린더에서 질소(N_2) 가스 또는 이산화탄소(CO_2)가 초임계 상태로 밀봉되어 일반 사출성형과 유사한 방법으로 충진, 냉각, 그리고 성형품을 취출한다.

MuCell 성형품과 일반 사출성형품의 차이는 없지만 제품을 들어보면 알 수 있다. 즉, MuCell 제품은 일반 성형 제품보다 약 12% 가볍다. 이런 차이의 이유를 알아보기 위해서 제품을 자르면 표면은 딘딘하지만 내부는 무수히 많은 매우 미세한 기공이 존재하며, 재료는 스폰지 (벌집 구조) 형상을 하고 있다.

In this manner, since the surface is solid but the inside has air bubbles, it is possible to achieve a lighter weight compared to the ordinary molded product which is completely solid. In addition, since this is a collection of very fine bubbles (with the bubble diameter being 50 micrometers or less), there is the unique feature that there is no marked decrease in the mechanical strength.

The features of this technology are the following.

1) Lightweight molded products can be obtained.
2) The molding cycle is made shorter, because the solidification time of the central part becomes shorter.
3) It is possible to set the screw heating temperature to a lower value, which results in energy saving.
4) It is possible to lower the viscosity of the plastic. Therefore, it is possible to fill at a low pressure.
5) The pressure maintenance time can be shortened.
6) It is possible to reduce warping and deformation substantially. The principle of operation is similar to gas assisted molding.
7) There is no difference in the outer appearance of the surface between the molded product and an ordinary molded product.
8) A wide range of plastics such as PS, ABS, PP, PA6, PPS, glass fiber reinforced plastic, etc., can be used.
9) It is possible to increase the number of products molded using the mold during its life.

As explained above, this new molding technology has technical content that needs to be mentioned particularly, and it appears that this technology is very highly likely to be applied in various fields from now on.

이러한 방식으로 표면은 고체이지만 내부는 기공이 있기 때문에 완전하게 고체인 통상의 성형 제품에 비해 더 가벼운 무게를 달성 할 수 있다. 또한, 이것은 매우 미세한 기공(기공 직경이 50 마이크로미터 이하)의 집합이기 때문에 기계적 강도의 현저한 감소가 없는 독특한 특징을 가지고 있다.

이 기술의 특징은 다음과 같다.

1) 경량의 성형품을 얻을 수 있다.
2) 중앙부의 응고시간이 짧아지기 때문에 성형 사이클이 단축된다.
3) 스크류 가열 온도를 낮게 설정하여 에너지를 절약 할 수 있다.
4) 플라스틱의 점도를 낮출 수 있다. 따라서 저압으로 충전이 가능하다.
5) 압력 유지시간을 단축 할 수 있다.
6) 휨 및 변형을 실질적으로 감소시키는 것이 가능하다. 작동 원리는 가스 사출성형과 유사하다.
7) 일반 성형품의 표면과 외관차이는 없다.
8) PS, ABS, PP, PA6, PPS, 유리섬유 강화 플라스틱 등과 같은 여러 종류의 플라스틱을 사용할 수 있다.
9) 수명 동안 금형을 사용하여 성형품 개수를 늘릴 수 있다.

위에서 설명한 바와 같이, 이 새로운 성형 기술은 특별히 언급 할 필요가 있는 기술적인 내용을 가지고 있으며, 이 기술은 지금부터 다양한 분야에 적용될 가능성이 매우 높다.

A license agreement is necessary in order to use this technology, and at present, it is necessary to enter into an implementation rights contract with Trexel via the manufacturers of the molding machine.

이 기술을 사용하려면 라이센스 계약이 필요하며, 현재 성형기 제조업체를 통해 Trexel과 계약을 체결해야한다.

J 002 Molds for Plastic Magnets

J-002 플라스틱 자석용 금형

Plastic magnets are molding materials produced by mixing magnetic powders in a plastic resin or an elastomer. Using injection molding, it is possible to manufacture magnetic molded products with a high degree of freedom in the shape.

Compared to sintered magnets, it is possible to produce products that are lightweight and have shapes with thin walls. However, the magnetic force is larger in sintered magnets.

The following types of magnetic powders are used.

플라스틱 자석은 자성 분말을 플라스틱 수지 또는 엘라스토머에 혼합하여 만든 성형품이다. 사출성형을 사용하여 형상의 자유도가 높은 자성체를 성형 할 수 있다.

소결 자석에 비해 가볍고 얇은 살 두께를 가진 제품을 생산할 수 있다. 그러나 자력은 소결 자석이 더 크다.

다음 유형의 자성 분말이 사용된다.

(1) Ferrite type

Barium ferrites	Strontium ferrites

(2) Rare earths type

Samarium	cobalt	
Samarium	iron	nitrogen
Neodymium	iron	Boron

(1) 페라이트 형

바륨 페라이트	스트론튬 페라이트

(2) 희토류

사마륨	코발트	
사마륨	철	질소
네오디뮴	철	붕소

Further polyamides (nylon) such as PA6, PA66, PA12, etc. are used as binders.

In some cases, PPS (polyphenylene sulfide) and PVC (polyvinyl chloride) are also used as binders.

In the case of plastic magnets, in some cases the devices for generating magnetic fields are incorporated in molds for giving polarities to the molded product inside the mold.

There are many methods for this such as having a structure in which permanent magnets are embedded inside the cavity, strong magnetic fields are generated using coils, or receiving the supply of magnetic fields from the molding machine, etc.

Therefore, for all components of molds, it is necessary to separate the use of magnetic materials (steel, nickel, etc.) and non-magnetic materials (stainless steel, copper alloys, etc.).

In addition, since the mold gets worn strongly due to magnetic powders, it is necessary to take measures to enhance the resistance to wear.

Since the molded products taken out stick to each other due to magnetic force, care should be taken in the handling of molded products after they are taken out (making suitable trays, etc.).

추가적으로 폴리아미드(나일론), 예컨대 PA6, PA66, PA12 등이 결합제로서 사용된다.

어떤 경우에는 PPS(폴리페닐렌설파이드) 및 PVC(폴리염화비닐)도 결합제로 사용된다.

플라스틱 자석의 경우, 금형 내부의 성형 제품에 극성을 부여하기 위한 자기장 발생 장치는 금형에 통합되는 경우도 있다.

이와 같은 방법으로는 캐비티 내에 영구 자석을 매립한 구조, 코일을 이용하여 강한 자력을 발생시키는 구조, 또는 성형기 등으로부터 자력을 공급받는 등의 많은 방법이 있다.

따라서 금형의 모든 부품에 대해 자성 재료(강철, 니켈 등)와 비자성 재료(스테인리스 강, 구리 합금 등)의 사용을 분리해야한다.

또한, 자성 분말에 의해 금형이 강하게 마모되기 때문에 내마모성을 높이는 대책을 강구 할 필요가 있다.

취출 된 성형품은 자력에 의해 서로 달라붙기 때문에 성형품의 취출에 주의를 기울여야한다(적절한 트레이 등을 만듦).

J 003 Molds for Medical Parts

Plastics are used a lot in medical parts. Injection molding is used widely as the method of their manufacture. Therefore, it can also be said that a substantial part of the medical parts are being manufactured using molds.

Some typical medical parts made of plastic are the following.

1) Cylinders of syringes (injection tube): PP, PE
2) Pistons of syringes: PP
3) Pipette tips: PP
4) Catheters: PVC, PC
5) Blood collection test tubes: PC
6) Petri dishes for culture growth: PS

The plastic materials used for medical parts, depending on the application, can only be used if they satisfy the quality standards of laws related to pharmaceuticals and drugs or of regulations of related governmental ministries or organizations. The use of these materials is limited to plastics that can withstand sterilization processing by irradiation with ultraviolet rays or gamma rays, and to plastics that have passed the clinical approval tests such as blood coagulation reactions or allergy reactions, etc., on human body.

In addition, since these devices are basically of a disposable type, the materials should also be those that do not destroy the environment when disposed off by burning.

I-003 의료 부품용 금형

플라스틱은 의료용 부품에 많이 사용된다. 사출성형은 의료용 부품을 생산하는 방법으로 널리 사용된다. 따라서 의료용 부품의 상당 부분이 금형을 사용하여 생산되고 있다고 할 수 있다.

플라스틱으로 만들어진 일부 전형적인 의료용 부품은 다음과 같다.

1) 주사기의 실린더(주입 튜브) : PP, PE
2) 주사기의 피스톤 : PP
3) 피펫 팁 : PP
4) 카테터 : PVC, PC
5) 혈액 수집 시험관 : PC
6) 배양 접시 : PS

응용 분야에 따라 의료용 부품에 사용되는 플라스틱 재료는 의약품 및 의약품 관련 법규 또는 관련 정부 부처 또는 기관의 품질 기준을 충족하는 경우에만 사용할 수 있다. 이 재료의 사용은 자외선이나 감마선 조사에 의한 멸균 처리에 견딜 수 있는 플라스틱 및 혈액 응고 반응이나 알레르기 반응 등 임상 승인 시험에 합격한 플라스틱에 한한다.

또한 이러한 장치는 기본적으로 일회용이므로 재료는 연소시켜 폐기할 때 환경을 파괴하지 않는 재료라야 한다.

It is recommended that molds are manufactured using materials that are resistant to rusting. Therefore, structures preventing corrosion of the cavity and core are used widely such as using stainless steel or ion plating coatings, hard chromium electroplating, etc. Further, valve gate molds are also used widely for molded products that are manufactured in very large quantities.

Since molded products become quality defects if burrs are present on the periphery of the molded product, precision products are used for the positioning guides on the movable side and fixed side, and also various techniques are adopted for preventing abnormal wear of core pins of various structures.

In order to stabilize the quality of molded products, since it is necessary to design delicately even the mechanisms for the temperature control of cavities and cores, it is necessary to pay attention to the structures of the cooling circuits and heat pipes. The cooling structure on the interior of the core becomes extremely important in the case of the tubes of syringes, etc.

In terms of quality control, removing any dirt or soot adhered to the surfaces of molds is also a very important point. Setting up of gas vents and forced suction structures are also being used.

금형은 부식에 강한 소재를 사용하여 생산하는 것이 좋다. 따라서 캐비티 및 코어의 부식을 방지하는 스테인리스 스틸 또는 이온 플레이팅 코팅, 경질 크롬 전기 도금 등이 널리 사용된다. 또한, 밸브 게이트 금형은 대량으로 생산되는 성형품에 널리 사용된다.

성형품의 외부에 버가 존재하면 성형품 불량이 되기 때문에 가동측 금형과 고정측 금형의 위치 결정 가이드에 정밀 부품을 사용하고, 각종 구조의 코어 핀의 이상 마모를 방지하기 위한 각종 기술을 채용하고 있다 .

성형품의 품질을 안정화시키기 위해서 캐비티 및 코어의 온도 제어 메커니즘을 정교하게 설계 할 필요가 있기 때문에 냉각 회로 및 히트 파이프의 구조에 주의를 기울여야한다. 코어 내부의 냉각 구조는 주사기 튜브 등의 경우 매우 중요하다.

품질관리 측면에서 금형 표면에 부착 된 먼지나 그을음을 제거하는 것도 매우 중요하다. 가스 배출구 및 강제 흡입 구조의 설치도 사용되고 있다.

J 004 Molds for Food Containers

A number of plastics are used for food containers. PET bottles, food cups, and food packaging are mostly made of plastics.
Quality control of food containers is necessary to ensure that they are hygienic and that they do not have any flash that can hurt fingers or lips so that people can feel safe about consuming food from the containers.
In addition, even quality defects such as holes or cracks are not allowed because bacteria can penetrate through them and cause the food to rot.

Some typical food containers are the following.

- Margarine container: PP
- Ice cream container: HIPS, HDPE
- Lactobacillus beverage (fermented milk drink) container: HIPS
- Pudding container: PS, PP
- Snacks container: HIPS, PP, HDPE
- Soft drinks bottle: PET

Depending on the usage of the food item, the container will have to have heat resistance (for foods that are heated and consumed), low temperature resistance (for foods consumed after refrigerating or freezing), or gas barrier ability (for food items that are to be stored so that they do not come into contact with oxygen).

I-004 식품 용기용 금형

많은 플라스틱이 식품 용기에 사용된다. 페트병, 식품 컵 및 식품 포장은 주로 플라스틱으로 만들어진다. 식품 용기의 품질관리는 위생적이며 손가락이나 입술에 해를 끼칠 수 있는 버가 없어야 사람들이 안심할 수 있다. 박테리아가 침투하여 음식물이 부패 할 수 있기 때문에 구멍이나 균열과 같은 품질 결함조차도 허용되지 않는다.

몇 가지 일반적인 식품 용기는 다음과 같다.

- 마가린 용기 : PP
- 아이스크림 용기 : HIPS, HDPE
- 락토 바실러스 음료(발효유 음료) 용기 : HIPS
- 푸딩 용기 : PS, PP
- 스낵 용기 : HIPS, PP, HDPE
- 소프트 드링크 병 : PET

식품의 사용에 따라 용기는 내열성(가열 및 소비되는 식품의 경우), 저온 내성(냉농 또는 냉농 후 소비되는 식품의 경우) 또는 가스 투과 방지 능력(식품의 경우 산소와 접촉하지 않도록 보관할 것) 등이 필요하다.

Since the sales of food containers is influenced greatly by the design (treatment for beautiful appearance) of the molded item, three dimensional curved shapes and designs are given importance. Therefore, three dimensional solid data becomes important in the design of the molded item and in the mold design. In addition, since there is the risk of flash being generated at the parting surface of the mold, it is necessary to give sufficient considerations to the parting surface.

Similarly, even the position and method of providing gates are also important.

It is necessary that the cavity and the core are made of a steel material having very good corrosion resistance. Also, since it is desirable that the mold is not coated with grease or oil for promoting sliding, the recomm ended mold structure is one which can operate safely without any lubrication.

Electrical motor driven type injection molding machine is preferable, and it is still better that the molding and machining is done in a clean room. Valve gates and hot runners can be used. If large volume production is expected to grow, the effect of eliminating scrap and high cycle speeds are extremely advantageous.

식품 용기의 판매는 성형품의 디자인(미려한 외관을 위한 처리)에 의해 크게 영향을 받기 때문에 3차원 곡선 형태 및 디자인이 중요하다. 따라서 3차원 솔리드 데이터는 성형품의 설계 및 금형설계에서 중요하다. 또한, 금형의 파팅면에서 플래시가 발생할 위험이 있기 때문에 금형의 파팅면에 대한 충분한 고려가 필요하다.

마찬가지로 게이트 위치와 방법도 중요하다.

캐비티 및 코어는 내식성이 우수한 강재로 제작 될 필요가 있다. 또한, 미끄럼을 촉진하기 위해서 금형에 그리스 또는 오일을 도포하지 않는 것이 바람직하기 때문에 윤활제 없이 안전하게 작동 할 수 있는 금형 구조를 권장한다.

전동 모터 구동식 사출성형기가 바람직하며, 성형 및 가공은 클린룸에서 이루어진다. 밸브 게이트 및 핫 런너를 사용할 수 있다. 대량 생산이 증가 할 것으로 예상되는 경우 스크랩 제거 및 고속도 사이클의 효과에 매우 유리한다.

플라스틱 사출금형 설계의 기본과 노하우

2019년 8월 10일 초판 인쇄
2019년 8월 20일 인쇄 발행

저 자 한성렬, 박태원
발 행 인 송기수
발 행 처 도서출판 위즈덤플
편 집 처 도서출판 위즈덤플
편 집 자 공예서
인 쇄 처 트윈벨미디어
등 록 번 호 제 2015-000009 호
I S B N 979-11-89342-07-4(93550)

주 소 서울 은평구 증산로 15길 69, 2층
전 화 02-976-7898, 02-3272-7898
팩 스 02-6468-7898
홈 페 이 지 gsintervision.co.kr
E-Mail gsinter7@gmail.com

정 가 25,000원